Jan Tromp

Gas-Phase Dissociation of Oligonucleotides

AF062630

Jan Tromp

Gas-Phase Dissociation of Oligonucleotides

Studied by Electrospray Ionization Tandem Mass Spectrometry

Südwestdeutscher Verlag für Hochschulschriften

Impressum / Imprint
Bibliografische Information der Deutschen Nationalbibliothek: Die Deutsche Nationalbibliothek verzeichnet diese Publikation in der Deutschen Nationalbibliografie; detaillierte bibliografische Daten sind im Internet über http://dnb.d-nb.de abrufbar.
Alle in diesem Buch genannten Marken und Produktnamen unterliegen warenzeichen-, marken- oder patentrechtlichem Schutz bzw. sind Warenzeichen oder eingetragene Warenzeichen der jeweiligen Inhaber. Die Wiedergabe von Marken, Produktnamen, Gebrauchsnamen, Handelsnamen, Warenbezeichnungen u.s.w. in diesem Werk berechtigt auch ohne besondere Kennzeichnung nicht zu der Annahme, dass solche Namen im Sinne der Warenzeichen- und Markenschutzgesetzgebung als frei zu betrachten wären und daher von jedermann benutzt werden dürften.

Bibliographic information published by the Deutsche Nationalbibliothek: The Deutsche Nationalbibliothek lists this publication in the Deutsche Nationalbibliografie; detailed bibliographic data are available in the Internet at http://dnb.d-nb.de.
Any brand names and product names mentioned in this book are subject to trademark, brand or patent protection and are trademarks or registered trademarks of their respective holders. The use of brand names, product names, common names, trade names, product descriptions etc. even without a particular marking in this work is in no way to be construed to mean that such names may be regarded as unrestricted in respect of trademark and brand protection legislation and could thus be used by anyone.

Verlag / Publisher:
Südwestdeutscher Verlag für Hochschulschriften
ist ein Imprint der / is a trademark of
OmniScriptum GmbH & Co. KG
Heinrich-Böcking-Str. 6-8, 66121 Saarbrücken, Deutschland / Germany
Email: info@svh-verlag.de

Herstellung: siehe letzte Seite /
Printed at: see last page
ISBN: 978-3-8381-0966-4

Zugl. / Approved by: Bern, University of Bern, Dissertation, 2007

Copyright © 2009 OmniScriptum GmbH & Co. KG
Alle Rechte vorbehalten. / All rights reserved. Saarbrücken 2009

Table of contents

I. Introduction

Abstract 5

Applications of oligonucleotides 6
    Antigene oligonucleotides 6
    Antisense oligonucleotides 7
    Aptamers 10
    Modifications in oligonucleotides 11

Oligonucleotides in mass spectrometry 13
    DNA dissociation 14
    RNA dissociation 18
    Sequencing with mass spectrometry 22
    Modified oligo(deoxy)ribonucleotides in MS 24

Aim of the work 26

II. Publications

Tromp et al., J. Am. Soc. Mass Spectrom., 2005
    Gas Phase Dissociation of Oligoribonucleotides and their
    Analogs Studied by Electrospray Ionization Tandem Mass
    Spectrometry   **29**

Monn et al., Chimia, 2005
    Mass Spectrometry of Oligonucleotides   **37**

Tromp et al., Rapid Commun. Mass Spectrom., 2006
    Electrospray ionization tandem mass spectrometry of
    biphenyl-modified oligo(deoxy)ribonucleotides **43**

Schürch et al., Nucleosides Nucleotides Nucleic Acids, 2007
    Mass Spectrometry of Oligonucleotides **51**

III. Discussion

Discussion **58**
    Dissociation products 58
    Local control 61
    2'-modifications 64
    Base modifications 66

Conclusions **69**

Outlook **70**

IV. References

References **72**

I. Introduction

Abstract

Oligonucleotides have great potential in diagnostic and therapeutic applications. Especially antisense oligonucleotides, which hybridize to complementary mRNA, thus inhibiting gene expression, are of great interest. The high specificity of oligonucleotides offers promise for the development of new drugs for protein induced diseases. Several modifications were introduced in oligonucleotides to improve nuclease resistance, binding affinity and specificity. The various unnatural oligonucleotides are not compatible with the classical sequencing techniques. Therefore, new methods for fast and unambiguous sequence verification and detailed structural characterization are needed. Mass spectrometry has the potential to achieve these requirements. However, fundamental mechanistic aspects of oligonucleotide dissociation have to be clarified first.

In this work unmodified DNA, RNA and several ribose or base modified oligonucleotides were analyzed to elucidate the gas-phase dissociation mechanism. Experiments with 2'-modified oligonucleotides were performed to investigate the reasons for the different gas-phase fragmentation behavior of DNA and RNA. It was shown that the electronegative substituent and the mobile proton of the hydroxyl group play a key role in the dissociation mechanism. Oligonucleotides with C-glycosidic bound biphenyl substituents were analyzed to investigate the influence of the nucleobase in gas-phase dissociation. The experiments confirm that the loss of the nucleobase is the initiation step of the DNA dissociation. In contrast what happens in DNA dissociation, c- and y-ions, the preferred fragments of RNA dissociation, were generated independently of nucleobase loss. A mechanism for RNA gas-phase dissociation, in accordance with all experimental data, is presented.

Applications of oligonucleotides

Oligonucleotides are used to control gene expression in different ways. For example antigene oligonucleotides inhibit transcription by binding to dsDNA forming a triplex, antisense oligonucleotides target a specific mRNA sequence resulting in the inhibition of translation and aptamers bind to proteins. The wide variety and the high specificity to their target make oligonucleotides very attractive as potential drugs.

Antigene oligonucleotides

Antigene approaches based on the binding of triplex-forming oligonucleotides (TFOs) to dsDNA resulting in transcriptional arrest. In 1957 Felsenfeld and Rich first described triple-stranded DNA and RNA structures [1,2]. In the late nineteen-eighties it was discovered that triplex-forming oligonucleotides can be used to detect a specific DNA sequence in a double helix [3,4]. The binding of the TFOs to the dsDNA is achieved by Hoogsteen, or reverse Hoogsteen, hydrogen bonding. The TFOs strand is located in the major groove. The parallel or pyrimidine binding motif is characterized by T.AT and C+.GC Hoogsteen triplets between a homo-pyrimidine TFO that is parallel to the purine strand of the dsDNA. Another stable DNA triple-helix motif with a purine rich TFO which is antiparallel to the purine strand of the dsDNA is the purine motif. This motif is stabilized by reverse Hoogsteen hydrogen bonding of A.AT, T.AT and G.GC triplets. The TFO strand lies in the major groove [5,6,7].

An advantage of the antigene approach is that there are only two copies of dsDNA per cell which have to be targeted and therefore a lower concentration of oligonucleotides should be needed to block transcription. Nevertheless, antigene technology is far less in scientific focus over recent years compared with antisense technology. The reasons for this are manifold.

Two problems are that the triple helices are considerably less stable than double helices and that the binding rate of the triplex-forming oligonucleotide to the dsDNA is slow [5]. In addition, there is the lack of affinity of the TFOs to their target DNA and the fact that the accessibility of recognition motifs of the dsDNA is uncertain. Any oligonucleotide which targets dsDNA also has first to penetrate the chromatin structure of the cell nucleus [8]. Furthermore, there are some standard problems of oligonucleotides in physiological conditions, such as the biostability and undesired non-specific binding.

An improvement of the antigene approach is made by modified oligonucleotides. In the last few years TFOs with increased affinity and sequence recognition properties have been investigated. It has been shown that TFOs have potential as future therapeutic tools in an antigene strategy and as tools in life science. Promising modifications are, for example, sugar modifications at the 2'-aminoethoxy or locked nucleic acid sugar units which significantly enhance the stability of the triple helix [6]. Great effort has also been made to find bases which can recognize all four base pairs in dsDNA to remove the drawback of the restriction of TFO binding to homopurine/homopyrimidine DNA target sequences [9].

Antisense oligonucleotides

Antisense oligonucleotides are designed to bind by Watson-Crick base pairing complementary to a specific mRNA sequence thereby inhibiting gene expression by translational arrest. Thus, they can be used to determine the function of a gene or as a therapeutic agent in new drugs. Zamecnik and Stephenson were the first researchers who recognized the potential of oligonucleotides complementary to target RNA [13,14]. Over the last decade an enormous effort has been made to develop antisense oligonucleotides for the treatment of cancer, autoimmune, cardiovascular and infectious diseases and several of these oligonucleotides are now in clinical trials [10,12].

Mechanisms for antisense oligonucleotide action

There is a huge variety of mechanisms in the antisense approach to inhibiting translation. They can be divided into antisense oligonucleotides, which recruit cellular enzymes which degrade the mRNA, and oligonucleotides, which deactivate the RNA by steric blocking without destroying the RNA target sequence [10,11,12]. The next part describes the most relevant antisense mechanisms.

Ribonuclease H

The oldest and still very widespread antisense method is based upon Ribonuclease H, a ubiquitous enzyme that cleaves RNA which is hybridized to DNA. Antisense oligonucleotides, which can recruit Ribonuclease H, bind to a specific target mRNA. Then the enzyme cleaves the mRNA and the antisense oligonucleotide is released. In this catalytic way a single antisense oligonucleotide can destroy several mRNA copies. The disadvantage of these antisense oligonucleotides is that most chemical modifications prevent Ribonuclease H from functioning. Only native DNA and modified oligonucleotides with similar backbones, such as phosphorothioate serve as efficient substrates for Ribonuclease H. However, the very efficient cleaving mechanism and the omnipresence of Ribonuclease H in cells together with quite attractive pharmacokinetic and toxicological properties of phosphorothioate nucleic acids enable RNase H based oligonucleotides to be used very often and very successfully.

Enzyme-mediated cleaving mechanisms

Other antisense oligonucleotides are able to recruit diverse cellular enzymes to destroy the RNA. RNase L, a ubiquitous endoribonuclease, is activated by 2'-5' linked oligoadenylates. To activate RNase L Silverman and Torrence developed antisense oligonucleotides, which are conjugated to such 2'-5-linked oligoadenylates, which recruit the enzyme [15]. RNase P, an enzyme which is involved in the biosynthesis of tRNA, can be recruited to cleave a target RNA by antisense oligonucleotides that imitate features from the tRNA structure [16,17].

An upcoming antisense mechanism is the RNA interference gene-silencing mechanism. Small interfering RNA (19-25 base pairs) are double stranded RNA molecules which guide complementary mRNA to a protein complex called an RNA-induced silencing complex (RISC). The RISC cleaves the mRNA and inhibits its translation into protein [18].

Ribozymes

Natural hammerhead ribozymes catalyze either their own cleavage or the cleavage of other RNA without the assistance of cellular nucleases. Synthetic antisense oligonucleotides take advantage of this property. They include a hammerhead catalytic site and are thereby able to cleave their specific target RNA sequence [19].

Steric blocking of splicing and translation initiation

There are antisense oligonucleotides which do not cleave the RNA function through the blocking of translation initiation or the modulation of splicing. These oligonucleotides must hybridize very strongly to their target. This is often achieved by diverse modifications of the backbone, the sugar unit or the nucleobase. Morpholino oligonucleotides [20,21,22,23], tricyclo-DNA [24], 2'-O-methylated oligonucleotides [25], and peptide nucleic acids [26,27] are some prominent modifications which have already successfully proven their ability to block splicing and translational initiation.

Translation initiation is mediated by translation factors which recognize the 5' cap structure and engage ribosomal subunits for the scanning and identification of the transcripts start codon [12]. Antisense oligonucleotides forming stable duplexes at this position can sterically shield the initiation codon thus blocking the translation mechanism.

Another mechanism not depending on cellular nucleases uses antisense oligonucleotides to modify splicing. Splicing, the excision of non-protein coding RNA sequences, is sequence specific. Consequently, oligonucleotides hybridizing to a splice site can prevent the required cleaving reactions by the spliceosome resulting in altered gene expression. An application of blocking the splicing mechanism is correcting splicing defects which are the reason for mutations in some diseases [28,29].

Aptamers

Oligonucleotides can fold into complex three-dimensional forms with a wide range of binding specificities. Therefore, not only are nucleic acids potential therapeutic targets as discussed above, but also proteins, peptides, carbohydrates and small organic molecules are possible targets. Oligonucleotides called aptamers can interfere with enzyme catalysis or specific protein-protein interactions [31]. Aptamers are able to bind to proteins with high affinity and specificity. They can be selected from a pool of random oligonucleotides (DNA, RNA, modified oligonucleotides) by SELEX (systematic evolution of ligands by exponential enrichment) [35]. The enormous libraries consist nowadays of more than 10^{15} different oligonucleotides. The ability of aptamers to bind with a high affinity and high specificity to protein targets, such as growth factors or proteins directly involved in the cell cycle progression, open a wide field of potential therapeutic approaches as was recently reviewed by Bunka and Stockley [36].

Aptamers can also be used as fundamental molecular recognition elements in biosensors. Different analytical formats have been reported for detecting diverse analytes through the use of aptamers [32,33,34].

Modifications in oligonucleotides

Unmodified DNA and RNA are not useful as therapeutic agents because they are not stable, because of enzymatic degradation under physiological conditions. This is the main reason why most oligonucleotides have a modified backbone or modified sugar units or modified nucleobases or a combination of several modifications. In addition, chemical modifications can solve some other problems arising in oligonucleotide applications. For oligonucleotides to be functional therapeutic agents they need the following properties. As mentioned above they must resist cellular nucleases. They also need a very high affinity to their target for a strong binding. Furthermore a high specificity is required so that there are as few as possible non-specific interactions. In addition, they must be toxicologically harmless and finally, the uptake of oligonucleotides into cells, in vitro delivery and technical feasibility are also challenges. Most of the properties oligonucleotides need can be enhanced by modifications. Therefore, an unbelievable variety of modifications have been developed and so only the most promising and widespread modifications will be introduced in the following part.

Backbone

The most common backbone modifications are the phosphorothioates and methyphosphonates. They are quite resistant to degradation by cellular nucleases and also serve as a very efficient substrate for mRNA degrading Ribonuclease H in antisense approaches. But phosphorothioate modifications reduce the binding affinity to the complementary mRNA and can bind strongly to proteins in a non-specific manner.

The fact that all modifications have not only various advantages, but also some negative aspects, often leads to a combination of different modifications to reach maximal hybridization stability, affinity and specificity.

The variety of backbone modifications is huge, ranging from little-modified phosphorothioates, methyphosphonates and phosphoramidates, through amine linked oligonucleotides to peptide nucleic acids (PNA) with a peptidic backbone. Therefore several reviews which cover in detail the many other types of backbone modifications are mentioned [5,37,38,39].

Sugar

The 2'-position of the ribose offers a width range of different modifications. RNA/RNA duplexes are generally more stable than sequence corresponding RNA/DNA duplexes [40]. Consequently, other 2'-modifications were synthesized to increase the stability. 2'-O-methly oligonucleotides, 2'-fluoro and 2'-amino analogs have an increased RNA binding affinity caused by the electronegative substituent. To generate nuclease resistant oligonucleotides, large 2'-O-alkyl moieties were inserted to increase resistance further, but often this led to a decrease in the RNA binding affinity [30,39,41].

Besides the modification of the 2'-substituent there are also other modifications of the ribose. At the 1' position the base could be in a α or β orientation. Furthermore the 4'-oxygen could be replaced by, for example, sulphur or carbon. There are also modifications replacing the ribose by six-membered rings, such as morpholinos, or even more complex locked nucleic acids (LNA) [5,6,9,23,39].

Nucleobase

Binding affinity and specificity can be increased by modification of the nucleobase. Because base modifications are often not sufficient for nuclease resistance, they were combined with backbone or sugar modifications. The variety of base modification is reviewed in [41,42,43].

Oligonucleotides in mass spectrometry

Because of the spread of oligonucleotides as therapeutic and diagnostic agents, oligonucleotides are also a focus of mass spectrometry. Soft ionization methods such as electrospray (ESI) and matrix-assisted laser desorption ionization (MALDI) enable the analysis of whole biomolecules. With a multiple-stage mass spectrometer more and more detailed structural analyses have been done together with the identification and sequencing of oligonucleotides which are always increasing in size.

The identification of small oligonucleotides is quite simple. In contrast the interpretation of product ion spectra of even relatively small oligonucleotides can be much more complex because of the variety of backbone cleavage positions and of nucleobase loss. McLuckey et al. proposed in 1992 a nomenclature which includes all possible fragmentation products [44]. The nomenclature was made analogous to the nomenclature scheme widely used for peptides and proteins and is today the standard oligonucleotide nomenclature (Figure 1).

The four possible cleavage sites along the phosphodiester backbone are indicated by the lower case a-d for fragments containing the 5' termini and w-z for those containing the 3' termini. The subscripts indicate the number of bases from the corresponding termini. Base position in the sequence is specified by a numerical subscript starting at the 5' end. A typical 5'-terminal fragment ion with nucleobase loss is a_3-B_3(G) in which G indicates the cleaved base [44].

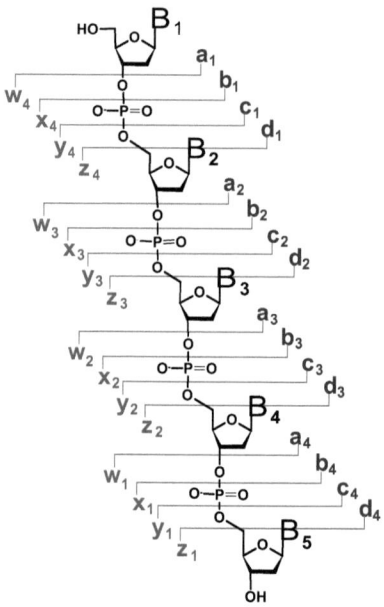

Figure 1. Oligonucleotide nomenclature.

DNA dissociation

Fragment ions

Oligodeoxyribonucleotides dissociation results in a great variety of different fragment ions. Product ion spectra show the cleavage of the phosphodiester backbone at every position and fragment ions with nucleobase loss are also often observed. The most intensive dissociation products in DNA dissociation are the w_n-ions generated by the cleavage of the 3'-C-O bond. The complementary $[a_n-B_n]$-ions and additional fragments due to backbone dissociation at other positions have reduced abundance.

DNA Dissociation mechanisms

In the nineteen-nineties several different research groups studied the gas-phase dissociation of oligodeoxyribonucleotides using MALDI or ESI. They found that DNA dissociation is dependent on nucleobase loss, ether negatively charged or neutral, and the following backbone dissociation generates predominantly w- and [a-B]-ions. Several diverse dissociation mechanisms were proposed to explain the formation of w- and [a-B]-ions. McLuckey et al., Rogers et al. and Barry et al. proposed promising mechanisms which begin with a 1,2 elimination of the nucleobase followed by different fragmentation mechanisms resulting in the cleavage of the 3'-C-O bond [45,46,47,48]. But these mechanisms could not explain the observed site-specific base loss or the absence of intermediates. Finally, H/D exchange experiments by Wan et al. proved that no 1,2 elimination occurs because the proposed mechanism involves the 2' non-exchangeable sugar proton [49]. The results demonstrate that base loss in deuterated oligonucleotides occurs only with deuterated bases. So all proposed mechanisms which include 1,2 elimination as an initiation step could be rejected. The H/D exchange experiments supported a mechanism proposed by Wang et al. in 1998 which does not include the 2' sugar proton and therefore is in agreement with the H/D exchange experiments [50].

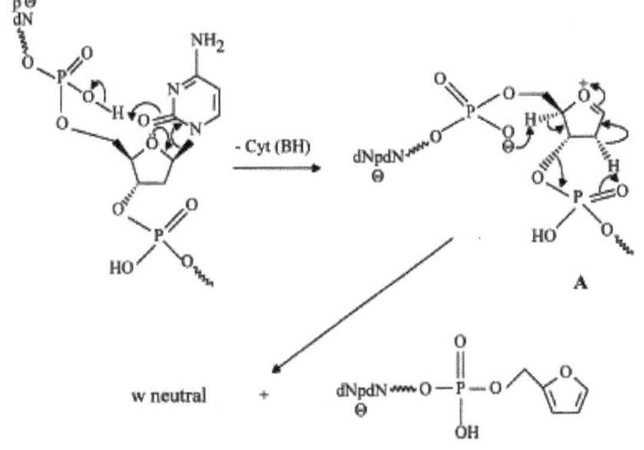

Figure 2. DNA dissociation mechanism proposed by Wang et al. [50].

The mechanism is induced by the transfer of the 5'-phosphate acidic proton of the adjoining phosphate to the nucleobase. This transfer initiates neutral base loss via a tripolar (zwitterion) intermediate. This loss thereby leads to the backbone cleavage to form the w- and [a-B]-ions [49]. The fragmentation mechanism proposed by Wang is accepted, but is not indisputable, because it can not explain fragmentation at the 5' termini. But until now no experiments have been able to reject the mechanism completely and so other explanations based on this mechanism were sought to explain terminal fragmentation.

Negative or neutral base loss

The first step in DNA dissociation is the loss of the nucleobase. This happens as a loss of either a neutral or a negatively-charged base [45,46,52]. For the understanding of the DNA dissociation mechanism the first step is of great importance. Therefore, several research groups are interested in the different mechanisms of base loss. The nucleobase loss is dependent on the one hand on the charge state of the parent ion and on the other hand on the identity of the nucleobase [48,51,52,53].

An oligonucleotide with a high ratio of the charge state compared to the phosphate groups tends to anionic base loss. The experimental trend of anionic base loss of the different bases is $A^- \gg G^- > T^- > C^-$ [51,52]. According to Pan et al. this fits well with the electrostatic potential calculations of the delocalization of electron density, with A^- being the most stabilized through delocalization [51].

In contrast, large oligonucleotides with a low charge state tend to neutral base loss. Neutral thymine (TH) is very infrequently cleaved as a neutral base. The other three bases showed no unambiguous trend. Several research groups obtained diverse trends depending on the charge state, sequence and size of the investigated oligonucleotide. Even the instrumentation methods seem to influence the preferences [48,54,55,56,57]. Another reason for the variable neutral base loss could be that a negatively-charged phosphate group enhances the proton affinity of the adjacent base. This leads to their preferred loss as a neutral base. Thus, the charge positions in oligonucleotides become interesting. In larger oligonucleotides tertiary structure also potentially influences neutral base loss by bring negative phosphate groups close to nucleobases by folding.

Sequence position of the negative charge

In negative ion mode the charges are located on the deprotonated phosphate groups. Collision induced dissociation (CID) of relatively short oligonucleotides allows the determination of the sequence and also the position of the negative charge.

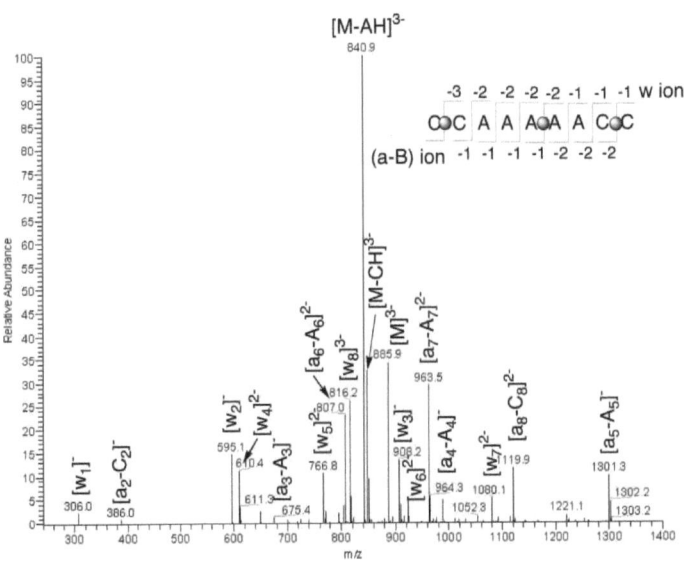

Figure 3. CID spectrum of $[M]^{3-}$. Gray circles indicate charge positions [51].

In the 9-mer shown, w_1 to w_3 are simple, w_4 to w_7 double and w_8 triple charged. The fact that there are no w-ions with different charge states suggests that the negative charge is not mobile during CID and stays on the phosphodiester group. The conclusion from the differently charged w-ions in regards to the dissociation mechanism (Figure 2) is that there has to be a charge on the first, fifth and eighth phosphodiester group from the 5' termini. Confirmation of the charge position is achieved by the complementary [a-B]-ions.

Positive ion mode

In solution, phosphate groups ($pK_a < 1$) are deprotonated leading to negatively-charged oligonucleotides. Therefore, oligonucleotides are typically analyzed using negative ion mode. However, against all expectations it is also possible to analyze oligonucleotides dissolved in water in the positive ion mode. With variation of the solvent and the pH, intensive signals can be achieved [58,59]. The nitrogens on the nucleobases are the most probable protonation sites [60]. But it is not clear if the phosphodiester groups are completely protonated or if their negative charge is compensated by more protonated bases building so-called zwitterions.

RNA dissociation

Compared with DNA studies, studies of RNA dissociation, sequencing and the influence of modifications are very rare. One reason for this is the delicate handling of RNA in the laboratory. RNase is found almost everywhere and degrades oligoribonucleotides very efficiently. A second point is that the 2'-hydroxy groups complicate the protection group chemistry in the synthesis for some modifications.

Fragment ions

In 1987 Cerny et al. first described the fragmentation products of RNA dissociation. Experiments with fast atom bombardment combined with tandem mass spectrometry (FAB-MS/MS) show RNA typical y-ions as a result of the backbone cleavage [63].

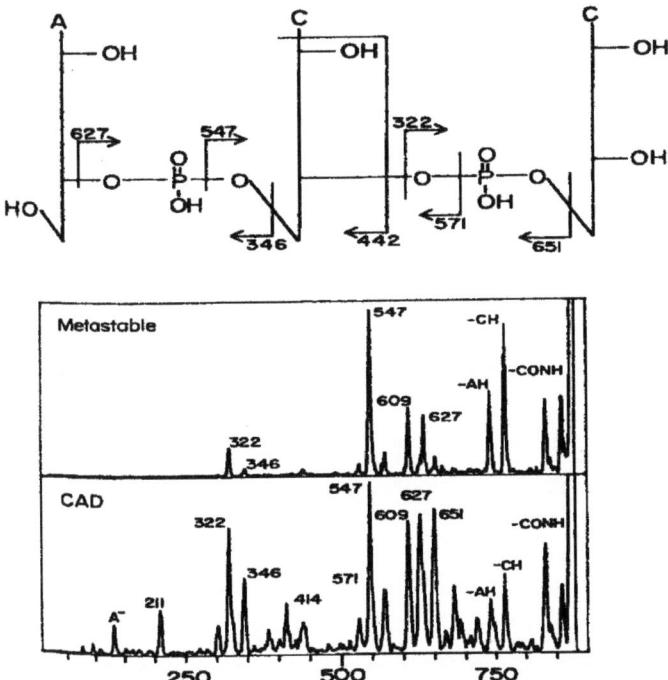

Figure 4. Scheme of the trinucleotide ACC. The backbone cleavage sites are indicated by arrows. Metastable decomposition spectrum and CAD spectrum of negative ions from [M-H]⁻ [63].

Several different fragment ions due to backbone cleavage were observed. The most abundant peak in both spectra is the y_2 ion with m/z of 547. Other intensive peaks correspond to w- and d-ions.

Collision induced dissociation (CID) of oligoribonucleotides results in complex product ion spectra. All fragment ions due to cleavages along the phosphodiester backbone were found, as was a loss of nucleobases. The predominant fragment ions were generally the c- and their complementary y-ions (Figure 1). In contrast to the all-dominant w-ions in DNA, the variety of c- and y-ions in the product ion spectra of RNA is not so noteworthy compared to the other fragmentation products.

RNA dissociation mechanisms

The difference in the fragmentation pattern of RNA compared to that of DNA allows for the conclusion that there is a changed dissociation mechanism. Because the 2'-hydroxy group is the only difference between RNA and DNA, this functional group must play a central role in the fragmentation mechanism. Mass spectrometric experiments have shown that RNA has an increased stability compared with DNA. This is caused by the electron withdrawing 2'-OH groups which have a stabilizing effect on the adjacent N-glycosidic bond [61,62].

Studies of RNA dissociation are very rare and experimental data specifically concerning the gas-phase dissociation mechanism of RNA are lacking completely. Cerny et al. first observed in FAB-MS/MS experiments the RNA typical c- and y-ions [63]. These complementary fragment ions are the result of the cleavage of the 5'-P-O bond. The same dissociation products were also described by Kirpekar et al. for positively-charged MADLI generated RNA ions and by Schürch et al. for negative ESI products [64,65].

A possible mechanism for gas-phase dissociation of RNA was proposed by Schürch et al. in 2002 [65].

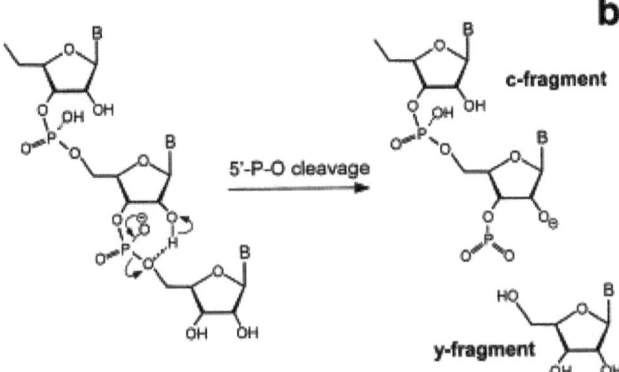

Figure 5. Proposed fragmentation mechanism of RNA dissociation resulting in c- and y-ions [65].

Initiation of backbone dissociation by the formation of an intramolecular cyclic transition state is proposed. The 2'-hydroxy proton is bridged to the 5'-phosphate oxygen. Following abstraction of the 2'-hydroxyl proton by the 5'-oxygen leads to the cleavage of the 5'-P-O bond generating c- and y-ions. Subsequent bond rearrangements lead to the formation of stable c-ions. The idea of an RNA dissociation mechanism, which involves the 2'-hydroxyl proton is promising. Therefore, different experiments were performed to investigate related dissociation mechanisms and to elucidate the details of backbone fragmentation.

In 2006 Anderson et al. published several fragmentation mechanisms for the different dissociation products generated in MALDI-MS [66]. They studied the mechanism of RNA and DNA dissociation by H/D-exchange in positive ion mode.

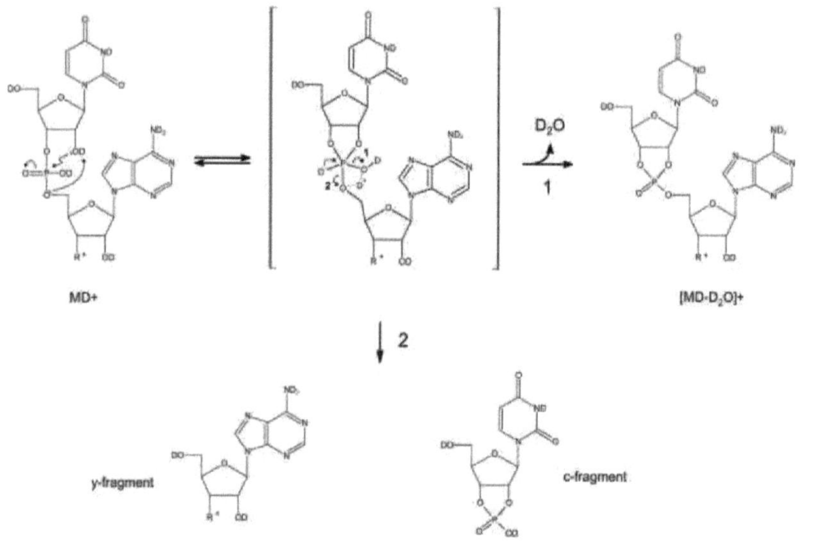

Figure 6. Proposed fragmentation mechanism for c- and y-ion formation from deuterated RNA [66].

The mechanism, initiated by an intramolecular nucleophilic attack of the 2'-hydroxl on the phosphor atom, yield a phosphorane transition state. The proposed initiation of the backbone dissociation is analog with the mechanism published by Tromp et al. (JASMS, Scheme 2, p. 35). The experiments also demonstrate that nucleobase loss generally does not precede cleavage of the 5'-P-O bond. This supports the conclusions of the experiments presented in this work.

Sequencing with mass spectrometry

In 1977 two revolutionary methods for sequencing oligonucleotides were published. On the one hand, there was chemical cleavage of Maxam and

Gilbert, which was the method of choice in the early nineteen-eighties [67]. On the other hand, there was the chain or dideoxy-termination method of Sanger, which is to this day the most widely used and efficient method for DNA sequencing [68].

DNA sequencing

In addition to the identification of oligonucleotides by molecular mass, MS/MS spectra offer a lot of information about base sequence. First experiments with DNA sequencing were restricted to small oligonucleotides [47,69,70]. Today sequence verification of nucleic acids up to 80-mers or even larger oligonucleotides is made by tandem mass spectrometry [71]. Oberacher et al. also developed automated *de novo* sequencing of small oligonucleotides (5 to 12-mer) by liquid chromatography MS/MS [72]. But in the near future these methods will not replace Sanger sequencing of natural DNA. The great potential of sequencing oligonucleotides with mass spectrometry lies in the analyses of modified oligonucleotides. Classical sequencing methods often fail in the presence of unnatural structural elements. Mass spectrometry can provide the high degree of structural information required and is capable of considerable automation. A current example is the detection of base substitution in PCR products [73].

RNA sequencing

To elucidate the RNA sequence, the oligoribonucleotides were transcribed to cDNA by reverse transcriptase and subsequently analyzed by the dideoxy-termination method [74]. Like DNA sequencing, sequencing of natural RNA is practicable. But the focus in research lies more in the development of mass spectrometric based methods which offer sensitive and efficient solutions to analyze, for example, posttranscriptionally modified nucleobases [75].

Modified oligo(deoxy)ribonucleotides in MS

Modified oligonucleotides are important compounds in current and future therapeutic applications. Because most modifications render the oligonucleotides immune to standard sequencing protocols, alternatives are needed. Mass spectrometry offers fast and reliable solutions for sequence verification and detailed analysis of complex structural problems.

There are two reasons why modified oligonucleotides are studied by mass spectrometry. On the one hand, one can identify modifications in therapeutic agents or natural occurring modifications. For example, Herdewijn and coworkers replaced the five-membered sugaring of natural oligonucleotides with a six-membered ring to make them stable against nucleases. For the sequence verification tandem mass spectrometry was used because classical sequencing methods do not function with these modified oligonucleotides [76]. Qui et al. used a mass spectrometric based method to detect ribose-methylated nucleotides, a common and conserved posttranscriptional modification of RNA [77]. Also nucleobase modifications can be identified by mass spectrometry because the modified bases influence the fragmentation pattern of oligonucleotides and this can therefore be used to locate them. For example, Pomerantz et al. designed a LC-ESI-MS/MS method for the detection of pseudouridine, a widespread posttranscriptional modified isomer of uridine [78]. On the other hand, modified oligonucleotides are in scientific focus because it is possible to investigate the influence of functional groups on the dissociation mechanism of oligonucleotides via a specific modification. Phosphodiester backbones were replaced by uncharged methylphosphonate linkages to elucidate the effect of backbone charge on the CID of oligonucleotides and identify the source of proton transfer in the DNA dissociation mechanism [79]. Oligonucleotides with a phosphorothioate backbone, a widely used modification in antisense approaches, show the same fragmentation behavior as natural oligonucleotides and can be easily sequenced by MS/MS. The influence of common 2' modifications on fragmentation patterns was analyzed by several research groups [80,81,82]. Electron-withdrawing groups, such as hydroxyl or fluoro groups, were of special interest because they stabilize the N-glycosidic bond. This results in a changed DNA fragmentation pattern because the initializing step of base loss is hindered.

Aim of the work

Mass spectrometry of oligonucleotides is nowadays primarily done for identification of oligonucleotides by their specific molecular weight. But there is a great unused potential of tandem mass spectrometric applications for detailed structural elucidation of oligonucleotides. Before mass spectrometric methods for sequence verification and identification and localization of structural modifications become standard, a better understanding of the mechanistic aspects of gas-phase dissociation of DNA, RNA and modified analogs is required.

The main focus in this work was to provide basic data about the gas-phase dissociation mechanism of oligoribonucleotides. In detail, the aim was to clarify the mechanistic differences in gas phase dissociation of DNA, RNA and other 2'-modified oligonucleotides. These differences are reflected in unequal fragmentation patterns. Furthermore the reason for the various abundances of complementary fragment ions generated by nucleobase loss in DNA was of great interest. Experiments with base modified oligonucleotides should clarify fundamental mechanistic aspects of nucleobase loss in the dissociation mechanism.

With a better understanding of the influence of specific functional groups on the dissociation mechanisms, interpretation of collision induced dissociation spectra should be simplified. Thereby, sequence verification and localization of the modification within the sequence should become possible.

II. Publications

Gas Phase Dissociation of Oligoribonucleotides and their Analogs Studied by Electrospray Ionization Tandem Mass Spectrometry

Jan M. Tromp and Stefan Schürch

Journal of the American Society for Mass Spectrometry
2005, 16, 1262-1268

The gas-phase dissociation mechanism of RNA proposed by Schürch et al. was the basis for this work (Figure 5). The proposed mechanism involves the 2'-hydroxyl proton and results in c- and y-ions.

In this work the first experimental data, which confirm the important role of the 2'-hydroxyl group in the RNA dissociation mechanism, are presented. The results show the stabilizing effect of an electronegative 2'-substituent on the N-glycosidic bond. In addition, experiments demonstrate that the availability of a proton in the proximity of the phosphate group is mandatory for the RNA typical backbone cleavage. Consequently, two possible dissociation mechanisms, both involving the 2'-hydroxyl proton, were proposed an investigated in detail. The mechanisms differ in the participation of the nucleobase. Therefore, experiments with base-modified oligonucleotides were performed, which demonstrate unambiguously the independence of the RNA dissociation mechanism on the nucleobase loss.

ARTICLES

Gas-Phase Dissociation of Oligoribonucleotides and their Analogs Studied by Electrospray Ionization Tandem Mass Spectrometry

Jan M. Tromp and Stefan Schürch
Department of Chemistry and Biochemistry, University of Bern, Bern, Switzerland

Oligoribonucleotides (RNA) and modified oligonucleotides were subjected to low-energy collision-induced dissociation in a hybrid quadrupole time-of-flight mass spectrometer to investigate their fragmentation pathways. Only very restricted data are available on gas-phase dissociation of oligonucleotides and their analogs and the fundamental mechanistic aspects still need to be defined to develop mass spectrometry-based protocols for sequence identification. Such methods are needed, because chemically modified oligonucleotides can not be submitted to standard sequencing protocols.
In contrast to the dissociation of DNA, dissociation of RNA was found to be independent of nucleobase loss and it is characterized by cleavage of the 5'-P—O bond, resulting in the formation of c- and their complementary y-type ions. To evaluate the influence of different 2'-substituents, several modified tetraribonucleotides were analyzed. Oligoribonucleotides incorporating a 2'-methoxy-ribose or a 2'-fluoro-ribose show fragmentation that does not exhibit any preferred dissociation pathway because all different types of fragment ions are generated with comparable abundance. To analyze the role of the nucleobases in the fragmentation of the phosphodiester backbone, an oligonucleotide lacking the nucleobase at one position has been studied. Experiments indicated that the dissociation mechanism of RNA is not influenced by the nucleobase, thus, supporting a mechanism where dissociation is initiated by formation of an intramolecular cyclic transition state with the 2'-hydroxyl proton bridged to the 5'-phosphate oxygen. (J Am Soc Mass Spectrom 2005, 16, 1262–1268) © 2005 American Society for Mass Spectrometry

Antisense oligonucleotides are nucleic acids of about 12 to 20 nucleobases, which are designed to hybridize to a complementary messenger RNA (mRNA) sequence, thus, inhibiting gene expression [1-3]. Therefore, they are of great interest in human cancer therapy and for diagnostic applications. Besides the high binding specificity of antisense oligonucleotides to their target mRNA, affinity, bioavailability, and biostability are of foremost importance. However, a major drawback of the application of unmodified phosphodiester DNA or RNA oligonucleotides is that these structures are subjected to rapid nuclease degradation under physiological conditions. To improve these factors, oligonucleotide analogs, exhibiting chemical modifications of the phosphodiester backbone, of the ribose, and, to a limited extent, also of the nucleobases, are evaluated.
Mechanisms of inhibition of gene expression by antisense oligonucleotides involve mRNA degrading enzymes such as RNase H, blockade of translation initiation, or modulation of splicing [4-7]. These mechanisms allow the application of new chemical modifications to increase the binding affinity and the nuclease resistance. Possible positions of the modifications are the phosphodiester backbone, the sugar unit and the nucleobases. Introduction of a phosphorothioate backbone increases nuclease resistance of antisense oligonucleotides and simultaneously serves as a very efficient substrate for mRNA degrading enzymes [6-9]. Unfortunately, such modification reduces the binding affinity to the complementary mRNA. Thus, further chemical modifications (e.g., derivatization of the ribose 2'-position by electronegative substituents) have been introduced to increase the affinity and to further enhance their nuclease resistance [10]. However, most of these highly modified oligonucleotide analogs do not support enzymatic activity and alternative mechanisms not relying on RNase H activity are required.
Apart from the development of appropriate synthetic methodologies, evaluation of oligonucleotide analogs is among the main focuses of antisense research. Evaluation of antisense oligonucleotides for therapeutic and diagnostic applications requires suitable analytical

Published online June 23, 2005
Address reprint requests to Dr. Stefan Schürch, Department of Chemistry and Biochemistry, University of Bern, Freiestrasse 3, CH-3012 Bern, Switzerland. E-mail: stefan.schuerch@ioc.unibe.ch

© 2005 American Society for Mass Spectrometry. Published by Elsevier Inc.
1044-0305/05/$30.00
doi:10.1016/j.jasms.2005.03.024

Received December 15, 2004
Revised March 22, 2005
Accepted March 23, 2005

tools for sequence determination and detailed structural elucidation. Because of the presence of unnatural structural elements, the classical sequencing techniques [11, 12] are likely to fail and alternative approaches for rapid and accurate sequencing of chemically modified oligonucleotides have to be developed. Tandem mass spectrometry is a highly attractive candidate for this task, because it is capable of providing the high degree of structural information required.

Within the past decade, detailed mechanistic data on the dissociation of unmodified oligodeoxyribonucleotides (DNA) in gas-phase have been provided. The loss of the nucleobase was found to be the initial dissociation step, leading to the subsequent cleavage of the 3'-C—O bond, thus, resulting in the formation of the generally abundant [a-B] and w ions. Several different mechanisms have been proposed and evaluated to explain the formation of the DNA typical [a-B] and w fragment ions [13-15] and the generally applied nomenclature of oligonucleotide fragments was proposed by McLuckey et al. in 1992 [16]. On the other hand, very limited data are available on the dissociation of RNA and oligonucleotide analogs.

Because of their use in antisense applications, methylphosphonate and phosphorothioate oligonucleotides are among the most widely studied analogs. Oligonucleotides with a methylphosphonate backbone have been investigated to study the effect of backbone charge on backbone dissociation and to trace the origin of the proton transferred to the nucleobase as the initial step of DNA dissociation [17, 18]. Oligonucleotides based on the phosphorothioate backbone are highly promising compounds for therapeutic antisense applications. The first of these antisense oligonucleotides was admitted to market in 1998 and further RNase H activating phosphorothioate oligonucleotides are currently in various stages of clinical trial [3, 19-21]. Investigation of the dissociation pattern of phosphorothioate, and phosphodiester oligodeoxyribonucleotides in gas-phase revealed that both backbone modifications yield the same types of fragment ions [22, 23].

Information on the fragmentation mechanism of RNA in gas-phase is rare. The few reports found indicate that dissociation of RNA clearly differs from the dissociation of DNA. The preferred cleavage site of RNA is the 5'-P—O bond, resulting in the formation of c- and y-type ions [24-26]. It has been shown that oligoribonucleotides possess higher gas-phase stability than oligodeoxyribonucleotides because of stabilization of the N-glycosidic bond by the 2'-OH substituent [10, 27]. Unlike dissociation of DNA, backbone dissociation of RNA is independent of nucleobase loss. Collision-induced dissociation (CID) of mixed sequence DNA/RNA oligonucleotides demonstrated that backbone cleavage is controlled locally and influenced by the type of adjacent ribose 2'-substituent. No evidence for any remotely located group affecting the dissociation was found and a fragmentation mechanism was proposed [26]. Furthermore, the fragmentation patterns of different 2'-ribose-modified oligonucleotides have been described without focusing on the mechanistic aspects of gas-phase dissociation [23, 28]. Reports specifically focusing on the dissociation mechanisms of 2'-modified oligonucleotides and their gas-phase stability are missing completely.

In this article we report on the dissociation of selected nucleobase- or 2'-ribose-modified oligonucleotides. To gain insight into the fragmentation mechanism of oligoribonucleotides and for better understanding of the influence of different 2'-substituents, a number of selectively modified oligonucleotides have been investigated by electrospray tandem mass spectrometry on a hybrid quadrupole time-of-flight mass spectrometer.

Experimental

Sample Preparation

RP-HPLC (reversed-phase high performance liquid chromatography) purified oligonucleotides were obtained from TriLink BioTechnologies (San Diego, CA) and Microsynth (Balgach, Switzerland). The lyophilized oligonucleotides were dissolved in a mixture of water/acetonitrile/triethylamine (49/49/2) resulting in a final oligonucleotide concentration of about 30 pmol/μL. Typically, a sample volume of 3 μL was used for each analysis.

Mass Spectrometry

All experiments were performed on an Applied Biosystems/MDS Sciex QStar Pulsar hybrid quadrupole time-of-flight mass spectrometer (Sciex, Concord, Canada), equipped with a nanoelectrospray ion source (Protana, Odense, Denmark). Oligonucleotides were analyzed in the negative-ion mode with a potential of −900 V applied to the nanospray needle. Nitrogen was used as the curtain gas.

Tandem mass spectrometric experiments were performed with precursor ions selected within a window of ±1.5 m/z units. CID was performed with collision energies in the range of 10-35 eV using nitrogen as the collision gas. The time-of-flight analyzer was tuned for an average mass resolving power of 10,000 (full width at half maximum) and calibrated externally using a mixture of cesium iodide and taurocholic acid. Calibration was performed before each series of experiments or at least once a day. The Applied Biosystems Analyst QS software package was used for data processing. Product ion spectra of tetranucleotides are based on the doubly deprotonated precursor ions.

Results and Discussion

A partial section of the product ion spectra of the tetranucleotides d(TATG) and UAUU is shown in Figure 1. It shows the unique fragment ion pattern of RNA, which clearly differs from the pattern of DNA. The

a

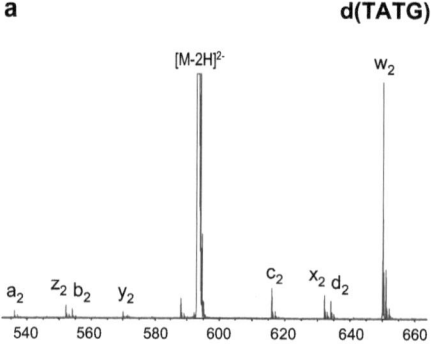

b

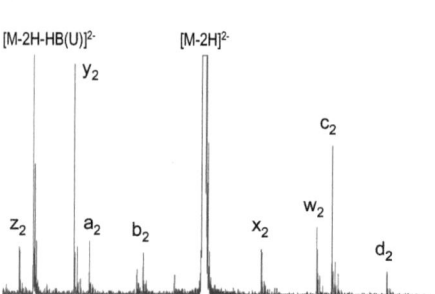

Figure 1. Comparison of the product ion spectra of DNA and RNA. (a) Product ion spectrum of the deoxyribonucleotide d(TATG), obtained by dissociation of the [M-2H]$^{2-}$ precursor ion with m/z 593.13. The abundant w_2 ion (m/z 650.13) is a result of the cleavage of the 3'-C—O bond, which is initiated by nucleobase loss. (b) The product ion spectrum of the [M-2H]$^{2-}$ precursor ion of the tetraribonucleotide UAUU (m/z 591.63) predominantly shows the RNA characteristic y_2 and c_2 ions with m/z 549.13 and 634.12, respectively. Other fragments, because of backbone cleavage, are of significantly lower intensity.

product ion spectrum of the doubly deprotonated tetradeoxyribonucleotide d(TATG) with m/z 593.13 is shown in Figure 1a. The spectrum is characterized by the abundant DNA typical w_1, w_2, and w_3 fragment ions with m/z 346.07, 650.13, and 963.20, respectively, generated by cleavage of the 3'-C—O bonds. The complementary [a-B]-type ions and additional fragments due to backbone cleavage at alternative positions and nucleobase loss are observed with reduced abundance. On the other hand, the product ion spectrum of the tetraribonucleotide UAUU ([M-2H]$^{2-}$, m/z 591.63) shows abundant y- and c-type ions (Figure 1b). Most of the theoretically possible fragment ions originating from backbone cleavage can be found in the spectrum. The y_2 ion (m/z 549.13) exhibits about a fourfold higher abundance than the other 3'-terminal ions (w_2, x_2, and z_2). Likewise, the abundance of the c_2 ion (m/z 634.12) is about three times higher, compared with the alternative 5'-terminal fragment ions (a_2, b_2, and d_2). The observed formation of c- and their complementary y-type ions confirms the 5'-P—O bond as the preferred cleavage site within the RNA backbone. RNA typical behavior was also found for the first and third phosphodiester group of the tetraribonucleotide (data not shown).

The different fragment ion patterns observed for DNA and RNA must be related to the presence of the altered 2'-substituent. The electronegative hydroxyl group of RNA plays a key role in the dissociation mechanism, because it effectively stabilizes the N-glycosidic bond. The inductive effect of such a strong electron-withdrawing group impedes formation of the carbocation at the 1'-position [10], thus, rendering nucleobase loss less prominent. Furthermore, the availability of the additional proton in the vicinity of the phosphodiester group provides alternative pathways for backbone cleavage.

To obtain mechanistic data on the gas-phase dissociation of RNA and to evaluate the influence of different 2'-substituents on the dissociation and stability of the N-glycosidic bond, a number of tetraribonucleotides bearing a single modified nucleotide have been studied. Modifications include 2'-methoxy and 2'-fluoro substituents at the 2'-position of the second nucleotide, and a deoxynucleotide lacking the nucleobase (dSpacer) as the third building block within the sequence (Scheme 1). Thus, the influence of the modifications on the second phosphodiester group can be observed. Tetramers were chosen for this study because they can be dissociated efficiently under low-energy CID conditions, and the length of the nucleotide sequences is still sufficient to study the effect of selected local modifications on backbone dissociation. Hydrogen/deuterium exchange experiments, which are often used for localizing the origin of protons transferred on dissociation reactions, do not provide conclusive information on the dissociation of RNA, because 2'-hydroxyl protons, remaining phosphate protons on the backbone, and amino protons of the nucleobases would be exchanged.

Compared with unmodified RNA, oligoribonucleotides with a 2'-methoxy-adenosine as the second nucleotide exhibits a different fragmentation behavior, as shown by the product ion spectrum of the doubly deprotonated U(2'OMeA)UU (m/z 598.60) in Figure 2a. Although the first and third phosphodiester groups dissociate into RNA typical y- and c-type ions, no evidence for a preferred dissociation pathway of the second phosphodiester group was found, and all of the theoretically possible fragment ions were observed with similar abundance. This experiment shows that the fragmentation mechanism of RNA is not exclusively influenced by the stabilization of the N-glycosidic bond

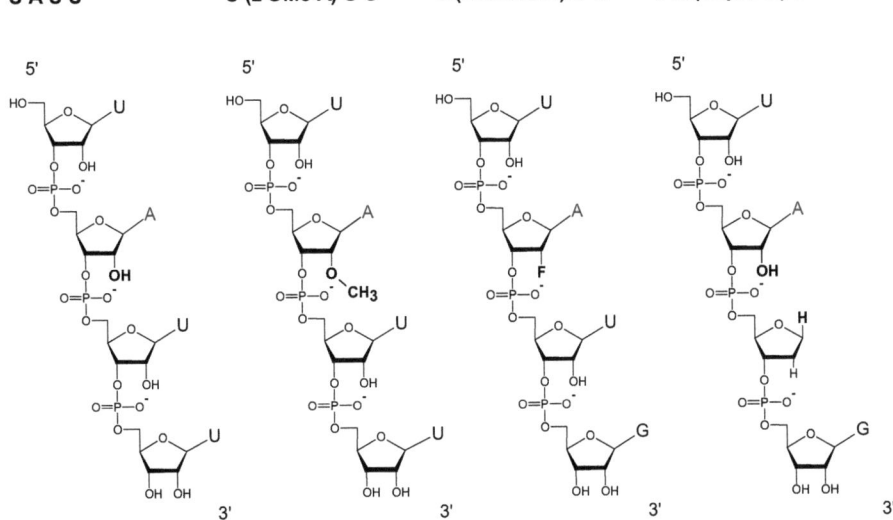

Scheme 1. Structures of the tetraribonucleotides investigated. All oligonucleotides are shown in completely deprotonated form. Unmodified RNA is shown on the left. Modifications are located between the second and the third building block and include 2′-methoxy and 2′-fluoro substituents and a deoxynucleotide lacking the nucleobase (dSpacer).

by the 2′-hydroxyl oxygen. The 2′-hydroxyl proton most likely contributes to the formation of y- and c-type ions as well. Otherwise, backbone dissociation next to 2′-methoxy-modified nucleotides would result in fragment ion patterns identical to those of unmodified nucleotides.

Figure 2b further shows the influence of an electronegative 2′-substituent on backbone dissociation. The product ion spectrum obtained by dissociation of doubly deprotonated U(2′FluoroA)UG (m/z 612.11), which bears a 2′-fluoro modification at the second nucleotide, is shown in Figure 2b. The stabilizing effect of the electronegative 2′-substiuent on the N-glycosidic bond reduces the rate of nucleobase loss and, consequently, the abundance of the corresponding fragment ions is decreased. Formation of the w_2 ion (m/z 668.10) even becomes less likely than the formation of the alternative x_2, y_2, and z_2 ions. Also, the complementary [a_2-B_2] ion (m/z 444.09) is of significantly reduced abundance, when compared with the dissociation pattern of an unmodified DNA. The 2′-fluoro nucleotide does not exhibit any preferred dissociation pathway, and the various types of fragment ions are of comparable abundance. The resulting fragment ion pattern is quite similar to the one observed for 2′-methoxy derivatives (Figure 2a). Although 2′-hydroxyl, 2′-methoxy, and 2′-fluoro substituents all stabilize the N-glycosidic bond, the fragmentation patterns of 2′-methoxy- and 2′-fluoro-modified tetranucleotides, which lack the exchangeable 2′-proton, clearly differ from the one of RNA. These experiments on oligonucleotides incorporating building blocks with modified 2′-substituents show the important role the 2′-hydroxyl group is playing in the dissociation of RNA. The electronegative 2′-hydroxyl group stabilizes the N-glycosidic bond and simultaneously provides a mobile proton.

A mechanism explaining backbone dissociation of RNA, which involves the 2′-hydroxyl proton, has been proposed previously [26]. It proposes initiation of backbone dissociation by formation of an intramolecular cyclic transition-state with the 2′-hydroxyl proton bridged to the 5′-phosphate oxygen. Abstraction of the 2′-hydroxyl proton by the 5′-oxygen leads to dissociation of the 5′-P—O bond and, finally, complementary c- and y-type fragment ions are released. Subsequent bond rearrangement leads to stable c-type fragment ions (Scheme 2a). The aforementioned results obtained from experiments with 2′-modified tetribonucleotides are consistent with this mechanism.

However, the possibility of an alternative gas-phase dissociation mechanism, similar to the one known for backbone cleavage of oligoribonucleotides in solution, still could not be excluded. The mechanism shown in Scheme 2b involves the 2′-hydroxyl proton and leads to the same RNA characteristic y- and c-type fragment ions as the mechanism shown in Scheme 2a. Although in gas-phase there is no free base present, which could act as a proton acceptor,

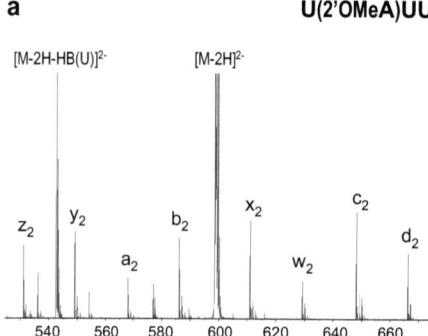

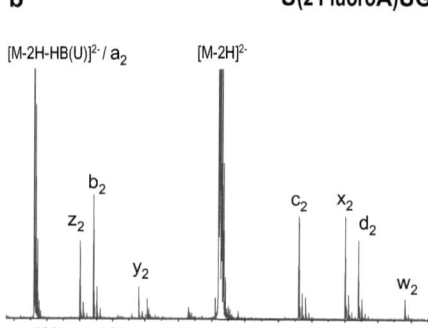

Figure 2. (a) Product ion spectrum of the doubly deprotonated modified tetraribonucleotide U(2'-OMeA)UU with m/z 598.60. The spectrum gives no evidence of a preferred dissociation pathway. (b) Product ion spectrum of doubly deprotonated U(2'-Fluoro)UG with m/z 612.11. The a_2 fragment ion (m/z 556.10) is overlapped by the very abundant [M-2H-HB(U)]$^{2-}$ ion.

deprotonation of the 2'-hydroxyl group could still be accomplished by intramolecular proton transfer to the adjacent 3'-nucleobase. Deprotonation of the 2'-hydroxyl group would be followed by transesterification, in which the 2'-oxygen participates as a nucleophile, generating a cyclic transition-state. Bond rearrangement and protonation of the leaving group (y ion) by the protonated nucleobase would finally result in cleavage of the phosphodiester backbone and in the formation of the c- and its complementary y-type fragment ions (Scheme 2b).

To verify such an alternative mechanism the tetranucleotide UA(dSpacer)G, incorporating a deoxy building block lacking the nucleobase (Scheme 1), has been subjected to CID. Incorporation of the dSpacer into the sequence enables evaluation of the role of the nucleobase in the dissociation of the phosphodiester backbone, because the dSpacer does not provide any nucleobase, which could potentially act as an acceptor for the 2'-hydroxyl proton. If the nucleobases were involved in the dissociation of RNA (e.g., as protonation sites), the presence of a building block lacking the nucleobase would block such mechanism. The altered fragmentation pathway should result in a product ion pattern that differs from the one observed for RNA. If, on the other hand, the nucleobase was not involved in the mechanism, dissociation of the phosphodiester backbone would not be affected by the dSpacer and c- and y-type ions would be formed predominantly.

Figure 3 shows an enlarged section of the product ion spectrum of the doubly deprotonated tetraribonucleotide UA(dSpacer)G with m/z 548.11. It represents the fragment ions generated by cleavage of the precursor between the second ribonucleotide (A) and the dSpacer. The RNA typical y_2 fragment ion (m/z 462.12) was observed with high abundance, whereas all alternative 3'-terminal fragment ions (x_2, z_2, and w_2) were of significantly decreased abundance, only reaching about 10-30% of the intensity of the y_2 ion. Likewise, c_2 (m/z 634.09) was the most abundant 5'-terminal fragment ion. The product ion spectrum of the unmodified oligoribonucleotide UAUU (Figure 1b) and the spectrum of the modified sequence incorporating the dSpacer (Figure 3) both show similar fragment ion patterns with the y- and c-type ions as the most abundant fragments. The fragment ion pattern is independent of the presence of the nucleobase. Consequently, the nucleobase does not have any significant influence on the dissociation mechanism of oligoribonucleotides. An alternative mechanism, in which the nucleobase plays a key role as acceptor of the 2'-hydroxyl proton, can be eliminated. These experiments clarify the role the nucleobases are playing in the fragmentation mechanism of oligoribonucleotides and confirm the previously proposed dissociation mechanism of RNA [26], which suggests backbone dissociation to be initiated by formation of an intramolecular cyclic transition-state with the 2'-hydroxyl proton bridged to the 5'-phosphate oxygen (Scheme 2a), while the adjacent nucleobase is not involved.

Conclusions

Investigation of oligoribonucleotides and their analogs by tandem mass spectrometry establishes the importance of the 2'-hydroxyl substituent as the structural key element in the fragmentation mechanism of RNA. In contrast to DNA, which predominantly fragment into [a-base]- and w-type ions, the presence of the 2'-hydroxyl group of RNA results in formation of abundant c- and y-type ions. The electron-withdrawing 2'-substituent induces a stabilizing effect on the N-glycosidic bond, thus, rendering nucleobase loss less prominent. In addition, the availability of a proton in

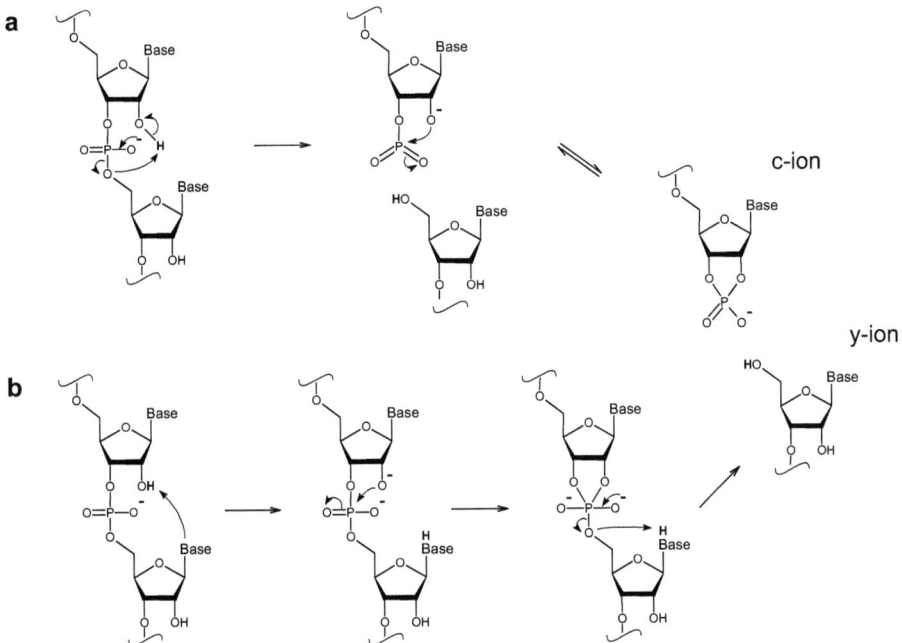

Scheme 2. Dissociation mechanisms of RNA. (a) Previously proposed mechanism describing the dissociation of RNA. (b) Alternative dissociation mechanism similar to the one known for backbone cleavage of oligoribonucleotides in solution. Both mechanisms include the 2′-hydroxyl proton and lead to the RNA-typical fragment ions. In contrast to the mechanism (a), the alternative mechanism (b) does involve the nucleobase.

the vicinity of the phosphate group is mandatory for the RNA typical backbone cleavage.

Experiments on oligoribonucleotides incorporating the dSpacer, a nucleotide lacking the nucleobase,

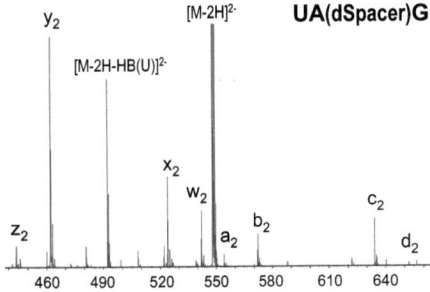

Figure 3. Product ion spectrum of the doubly deprotonated UA(dSpacer)G (m/z 548.11), which incorporates a building block lacking the nucleobase. The spectrum predominantly shows the RNA characteristic y_2 ion with m/z 462.12. Other fragments indicate backbone cleavage at alternative positions.

showed that gas-phase dissociation of the RNA backbone is not influenced by the nucleobases. Therefore, a mechanism similar to the one known for backbone cleavage of oligoribonucleotides in solution can be eliminated. The results support the mechanism in which backbone dissociation is initiated by formation of an intramolecular cyclic transition-state with the 2′-hydroxyl proton bridged to the 5′-phosphate oxygen.

Results provide fundamental information on the gas-phase dissociation of oligoribonucleotides. This information is needed to provide the basis for reliable tandem mass spectrometric sequencing methods, which will greatly facilitate the identification and verification of oligonucleotides and their analogs used for therapeutic applications, as genetic probes or as primers for amplification.

Acknowledgments

The authors thank the Swiss National Science Foundation and the University of Bern for generous support of this work.

References

1. Zamecnik, P. C.; Stephenson, M. L. Inhibition of Rous Sarcoma Virus Replication and Cell Transformation by a Specific Oligodeoxynucleotide. *Proc. Natl. Acad. Sci. U.S.A.* **1978**, *75*, 280–284.
2. Stephenson, M. L.; Zamecnik, P. C. Inhibition of Rous sarcoma Viral RNA Translation by a Specific Oligodeoxyribonucleotide. *Proc. Natl. Acad. Sci. U.S.A.* **1978**, *75*, 285–288.
3. Leumann, C. J. DNA Analogues: From Supramolecular Principles to Biological Properties. *Bioorg. Med. Chem.* **2002**, *10*, 841–854.
4. Baker, B. F.; Monia, B. P. Novel Mechanisms for Antisense-Mediated Regulation of Gene Expression. *Biochim. Biophys. Acta* **1999**, *1489*, 3–18.
5. Goodchild, J. Oligonucleotide Therapeutics: 25 Years Agrowing. *Curr. Opin. Mol. Ther.* **2004**, *6*, 120–128.
6. Altmann, K. -H.; Dean, N. M.; Fabbro, D.; Freier, S. M.; Geiger, T.; Häner, R.; Hüsken, D.; Martin, P.; Monia, B. P.; Müller, M.; Natt, F.; Nicklin, P.; Phillips, J.; Piels, U.; Sasmor, H.; Moser, H. E. Second Generation of Antisense Oligonucleotides: From Nuclease Resistance to Biological Efficacy in Animals. *Chimia.* **1996**, *50*, 168–176.
7. Myers, K. J.; Dean, N. M. Sensible Use of Antisense: How to Use Oligonucleotides as Research Tools. *TIPS* **2000**, *21*, 19–23.
8. Campbell, J. M.; Bacon, T. A.; Wickstrom, E. Oligodeoxynucleoside Phosphorothioate Stability in Subcellular Extracts, Culture Media, Sera and Cerebrospinal Fluid. *J. Biochem. Biophys. Methods* **1990**, *20*, 259–267.
9. Stein, C. A.; Tonkinson, J. L.; Yakubov, L. Phosphorothioate Oligodeoxynucleotides-Anti-Sense Inhibitors of Gene Expression? *Pharmacol. Ther.* **1991**, *52*, 365–384.
10. Tang, W.; Zhu, L.; Smith, L. M. Controlling DNA Fragmentation in MALDI-MS by Chemical Modification. *Anal. Chem.* **1997**, *69*, 302–312.
11. Sanger, F.; Nicklen, S.; Coulson, A. R. DNA Sequencing with Chain-Terminating Inhibitors. *Proc. Natl. Acad. Sci. U.S.A.* **1977**, *74*, 5463–5467.
12. Dovichi, N. J.; Zhang, J. How capillary electrophoresis sequenced the human genome. *Angew. Chem. Int. Ed.* **2000**, *39*, 4463–4468.
13. Wan, K. X.; Gross, J.; Hillenkamp, F.; Gross, M. L. Fragmentation Mechanisms of Oligodeoxynucleotides Studied by H/D Exchange and Electrospray Ionization Tandem Mass Spectrometry. *J. Am. Soc. Mass Spectrom.* **2001**, *12*, 193–205.
14. Wang, Z.; Wan, K. X.; Ramanathan, R.; Taylor, . John S.; Gross, M. L. Structure and Fragmentation Mechanisms of Isomeric T-Rich Oligodeoxynucleotides: A Comparison of Four Tandem Mass Spectrometric Methods. *J. Am. Soc. Mass Spectrom.* **1998**, *9*, 683–691.
15. McLuckey, S. A.; Habibi-Goudarzi, S. Decompositions of Multiply Charged Oligonucleotide Anions. *J. Am. Chem. Soc.* **1993**, *115*, 12085–12095.
16. McLuckey, S. A.; Van Berkel, G. J.; Glish, G. L. Tandem Mass Spectrometry of Small, Multiply Charged Oligonucleotides. *J. Am. Soc. Mass Spectrom.* **1992**, *3*, 60–70.
17. Bartlett, M. G.; McCloskey, J. A.; Manalili, S.; Griffey, R. H. The Effect of Backbone Charge on the Collision-Induced Dissociation of Oligonucleotides. *J. Mass Spectrom.* **1996**, *31*, 1277–1283.
18. Wan, K. X.; Gross, M. L. Fragmentation Mechanisms of Oligodeoxynucleotides: Effects of Replacing Phosphates with Methylphosphonates and Thymines with Other Bases in T-Rich Sequences. *J. Am. Soc. Mass Spectrom.* **2001**, *12*, 580–589.
19. Yacyshyn, B. R.; Chey, W. Y.; Goff, J.; Salzberg, B.; Baerg, R.; Buchman, A. L.; Tami, J.; Yu, R.; Gibiansky, E.; Shanahan, W. R. Double Blind, Placebo Controlled Trial of the Remission Inducing and Steroid Sparing Properties of an ICAM-1 Antisense Oligonucleotide, Alicaforsen (ISIS 2302), in Active Steroid Dependent Crohn's Disease. *Gut* **2002**, *51*, 30–36.
20. Sewell, K. L.; Geary, R. S.; Baker, B. F.; Glover, J. M.; Mant, T. G.; Yu, R.; Tami, J. A.; Dorr, A. Phase I Trial of ISIS 104838, a 2′-Methoxyethyl Modified Antisense Oligonucleotide Targeting Tumor Necrosis Factor-a. *JPET* **2002**, *303*, 1334–1343.
21. Herbst, R. S.; Frankel, S. R. Oblimersen Sodium (Genasense bcl-2 Antisense Oligonucleotide): A Rational Therapeutic to Enhance Apoptosis in Therapy of Lung Cancer. *Clin. Cancer Res.* **2004**, *10*, 4245–4248.
22. Wang, B. H.; Hopkins, C. E.; Belenky, A. B.; Cohen, A. S. Sequencing of Modified Oligonucleotides Using In-Source Fragmentation and Delayed Pulsed Ion Extraction Matrix-Assisted Laser Desorption Ionization Time-of-Flight Mass Spectrometry. *Int. J. Mass Spectrom. Ion Proccess* **1997**, *169/170*, 331–350.
23. Sannes-Lowery, K. A.; Hofstadler, S. A. Sequence Confirmation of Modified Oligonucleotides Using IRMPD in the External Ion Reservoir of an Electrospray Ionization Fourier Transform Ion Cyclotron Mass Spectrometer. *J. Am. Soc. Mass Spectrom.* **2003**, *14*, 825–833.
24. Cerny, R. L.; Tomer, K. B.; Gross, M. L.; Grotjahn, L. Fast Atom Bombardment Combined with Tandem Mass Spectrometry for Determining Structures of Small Oligonucleotides. *Anal. Biochem.* **1987**, *165*, 175–182.
25. Kirpekar, F.; Krogh, T. N. RNA Fragmentation Studied in a Matrix-Assisted Laser Desorption/Ionisation Tandem Quadrupole/Orthogonal Time-of-Flight Mass Spectrometer. *Rapid Commun. Mass Spectrom.* **2001**, *15*, 8–14.
26. Schürch, S.; Bernal-Mendez, E.; Leumann, C. J. Electrospray Tandem Mass Spectrometry of Mixed-Sequence RNA/DNA Oligonucleotides. *J. Am. Soc. Mass Spectrom.* **2002**, *13*, 936–945.
27. Nordhoff, E.; Cramer, R.; Karas, M.; Hillenkamp, F.; Kirpekar, F.; Kristiansen, K.; Roepstorff, P. Ion Stability of Nucleic Acids in Infrared Matrix-Assisted Laser Desorption/Ionization Mass Spectrometry. *Nucleic Acids Res.* **1993**, *21*, 3347–3357.
28. Ni, J.; Pomerantz, S. C.; Rozenski, J.; Zhang, Y.; McCloskey, J. A. Interpretation of Oligonucleotide Mass Spectra for Determination of Sequence Using Electrospray Ionization and Tandem Mass Spectrometry. *Anal. Chem.* **1996**, *68*, 1989–1999.

Mass Spectrometry of Oligonucleotides

Selina T. M. Monn, Jan M. Tromp and Stefan Schürch

Chimia
2005, 59, 822-825

This article is a short review based on the previous publications of our research group. It provides information about the elucidation of the gas-phase dissociation mechanism of RNA and shows an example of the characterization of metal-oligonucleotide complexes.

The results demonstrate the great potential of tandem mass spectrometric applications for structural elucidation of modified oligonucleotides and metal-oligonucleotide complexes.

Mass Spectrometry of Oligonucleotides

Selina T.M. Monn, Jan M. Tromp, and Stefan Schürch*

Abstract: Within the past decade mass spectrometry has undoubtedly consolidated its role as a major player in modern bioanalysis. Mass spectrometry of oligonucleotides is an evolving field that has not reached its culmination point yet, as many aspects of oligonucleotide dissociation in gas phase still need to be clarified. Research in our group is focused on the elucidation of the dissociation mechanisms of oligonucleotides and their analogs and on the characterization of metal–oligonucleotide complexes by tandem mass spectrometry in order to gain structural information. Here, we present the dissociation mechanism of RNA and an example for the characterization of metal–oligonucleotide complexes.

Keywords: Collision-induced dissociation · Iron(III) · Metal–oligonucleotide complexes · RNA · Tandem mass spectrometry

Introduction

Modified oligonucleotides play an important role in many disciplines of life sciences and medicine. Antisense oligonucleotides find application in human therapy or as genetic probes for diagnostic purposes in molecular biology. Aptamers exhibiting catalytic activity towards specific protein targets are another family of nucleic acid based therapeutics. Besides the enhancement of binding properties and biostability by the introduction of structural modifications, the physicochemical properties of oligonucleotides are influenced by incorporation of metal ions. Metal ions are important cofactors stabilizing oligonucleotide folding whereas certain organo-metallic compounds act as potent anti-tumor agents. Specific binding of metal ions to biomolecules bears a great potential for analytical purposes, as it affects the gas-phase behavior of oligonucleotides, thus providing complementary tandem mass spectrometric data.

Expanding research in the field of modified oligonucleotides demands suitable analytical tools for size and purity verification of known compounds and accurate structure elucidation of unknowns. There is a need to characterize the types and sites of modifications in oligonucleotides and to identify and sequence selected candidates originating from combinatorial syntheses.

Sequence Determination of Oligoribonucleotides (RNA) by Tandem Mass Spectrometry

Due to the unnatural structural elements present in antisense oligonucleotides, the classical techniques for sequence determination are likely to fail and alternative analytical tools are needed. Tandem mass spectrometry is a highly promising candidate for this task, as the method provides detailed structural information due to the structure-specific fragment ions formed upon collision-induced dissociation (CID), while requiring minute quantities of sample only.

In contrast to the automated mass spectrometric analysis of peptides and proteins, which is routinely applied in many laboratories, sequence determination of oligonucleotides by tandem mass spectrometry is hardly employed. When we started to address this topic a few years ago, we faced highly complex analytical data generated by tandem mass spectrometry of even short nucleotide sequences. The structural variations present in antisense oligonucleotides affect the dissociation mechanism along with the cleavage sites and consequently render the interpretation of analytical data difficult. To establish a reliable protocol for oligonucleotide sequencing based on tandem mass spectrometry, the fundamental mechanistic aspects of oligonucleotide dissociation in gas-phase have to be elucidated in detail.

While dissociation of oligodeoxyribonucleotides (DNA) in gas-phase has been investigated by several groups over the past decade [1–3], hardly any conclusive data on the dissociation of oligoribonucleotides (RNA) and modified oligonucleotides are found in literature. In contrast to published data on DNA, which shows that DNA predominantly dissociates into [a-B]- and w-ions due to cleavage of the 3'-P-O bond, we found the preferred cleavage site of the RNA backbone to be the 5'-P-O bond, resulting in abundant c- and y-ions [4]. Examples of DNA and RNA typical product ion spectra, along with the nomenclature of the fragment ions, are shown in Fig. 1. Furthermore, our data demonstrated that backbone dissociation of mixed-sequence DNA/RNA-oligomers is locally controlled and influenced by the adjacent ribose 2'-substituent only. Since the ribose 2'-substituent is the only structural variation between the DNA and RNA sequences investigated, it must be responsible for the altered dissociation behavior of the two types of oligonucleotides.

Correspondence: Dr. S. Schürch
Department of Chemistry and Biochemistry
University of Bern
Freiestrasse 3
CH-3012 Bern
Tel.: +41 31 631 43 89
Fax: +41 31 631 34 21
E-Mail: stefan.schuerch@ioc.unibe.ch
http://www.dcb.unibe.ch/groups/schuerch/

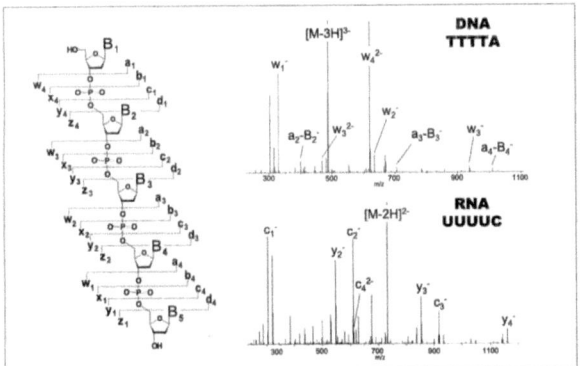

Fig. 1. Product ion spectra of d(TTTTA) and UUUUC. The spectrum of DNA is characterized by [a-B]- and w-fragment ions, whereas RNA primarily dissociates into c- and y-ions. The scheme shows the nomenclature of oligonucleotide fragments, as proposed by McLuckey et al. [1].

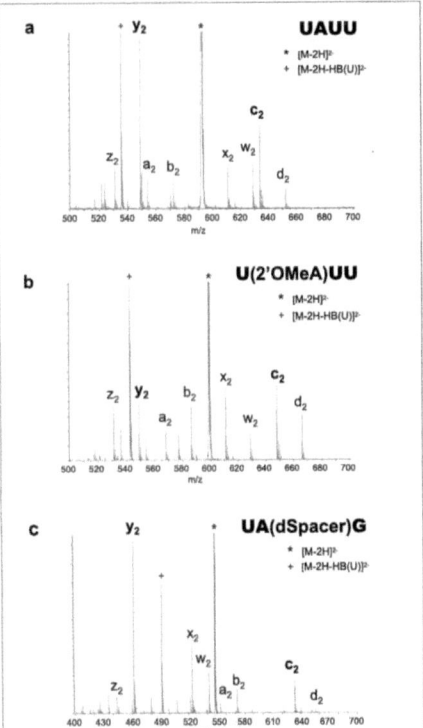

The mechanism of RNA dissociation in gas-phase was elucidated by studying the influence of various 2'-substituents and nucleobases on backbone cleavage, as both groups can potentially participate in the proton transfer that is involved in the dissociation mechanism of RNA. Fig. 2a shows a section of the product ion spectrum of the doubly deprotonated tetraribonucleotide UAUU (m/z 591.63). The abundant c- and y-ions confirm the 5'-P-O bond as the preferred cleavage site of the RNA backbone. The picture changes when the 2'-hydroxyl group is replaced by a methoxy substituent. As demonstrated by the product ion spectrum of U(2'OMeA)UU ([M-2H]$^{2-}$, m/z 598.60), the lack of the transferable proton inhibits the RNA typical dissociation pathway, resulting in all of the theoretically possible fragment ions to be generated with similar abundance (Fig. 2b). Experiments on oligoribonucleotides incorporating 2'-modifications demonstrated, that the availability of a transferable 2'-proton in the vicinity of the phosphodiester group contributes to the formation of the RNA typical c- and y-ions. Such behavior of RNA is opposed to the backbone dissociation of DNA, which is initiated by protonation and loss of the nucleobase.

We propose a gas-phase dissociation mechanism, where backbone cleavage is initiated by formation of an intramolecular cyclic transition state with the 2'-hydroxyl proton bridged to the 5'-phosphate oxygen [5]. Transfer of the 2'-hydroxyl proton to the 5'-oxygen leads to dissociation of the 5'-P-O bond and finally, complementary c- and y-ions are released. Subsequent bond rearrangement leads to stable c-ions (Fig. 3a). A dissociation mechanism similar to the one known for backbone dissociation of oligoribonucleotides in solution (Fig. 3b) might also be feasible in gas-phase, as it involves the 2'-hydroxy proton and leads to the same RNA typical c- and y-ions as well. To evaluate such alternative mechanism, the tetranucleotide UA(dSpacer)G, incorporating a building block lacking the nucleobase, has been subjected to MS/MS analysis. The modification allows the evaluation of the role of the nucleobase in the dissociation mechanism, since there is no nucleobase which could potentially be involved in a proton transfer. A section of the product ion spectrum of UA(dSpacer)G, reflecting the fragment ions generated by cleavage of the

Fig. 2. Product ion spectra of RNA and modified tetraribonucleotides. The [M-2H]$^{2-}$ ion was selected as the precursor for collision-induced dissociation. a) Spectrum of the unmodified tetraribonucleotide UAUU, showing the c- and y-ions as the most abundant fragments. b) Spectrum of U(2'OMeA)UU, which lacks the transferable 2'-proton of adenosine. The absence of any preferred type of fragment ion points out the key role of the 2'-hydroxy proton of RNA. c) Spectrum of UA(dSpacer)G, incorporating an abasic site as the third building block. Abundant c_2- and y_2-fragment ions are generated, despite the lack of the adjacent nucleobase.

phosphodiester adjacent to the dSpacer, is shown in Fig. 2c. The y_3-fragment ion is observed with more than threefold abundance of the other 3'-terminal fragment ions and the c_2-ion is the most abundant 5'-terminal one. The observed RNA typical fragment ion pattern demonstrates that the nucleobase is not involved in the dissociation mechanism of the phosphodiester backbone. The experiment confirms the dissociation mechanism given in Fig. 3a and an alternative gas-phase dissociation mechanism similar the one known for cleavage of oligoribonucleotides in solution can be rejected.

Tandem mass spectrometry does provide the accuracy, sensitivity and speed needed for characterization of this biologically significant class of compounds. However, increased understanding of the gas-phase behavior of natural and modified oligonucleotides is needed to take full advantage of the capabilities of modern analytical instrumentation for structure elucidation.

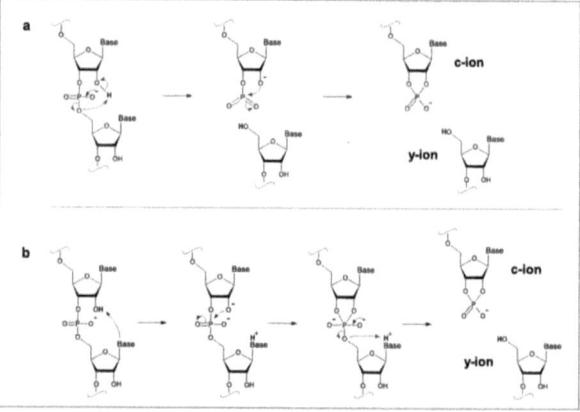

Fig. 3. a) Proposed mechanism for the gas-phase dissociation of RNA, resulting in c- and y-fragment ions. b) The alternative dissociation mechanism involving the nucleobase as a proton acceptor could be rejected.

Mass Spectrometry of Metal–Oligodeoxynucleotide Complexes

Exploration of metal–oligodeoxynucleotide complexes by tandem mass spectrometry is a topic that is rarely investigated but which holds great potential for future applications [6]. Pentadeoxynucleotides have been chosen as model compounds to elucidate the fragmentation pattern of metal–oligonucleotide complexes [7]. While oligonucleotides are commonly analyzed in the negative ion mode, charge compensation caused by complexation of metal cations renders analysis in the positive ion mode more favorable. Our studies demonstrated that the fragmentation of uncomplexed protonated oligodeoxynucleotides does not differ significantly from the fragmentation pattern of deprotonated oligodeoxynucleotides, mainly resulting in the w- and (a-B)-fragment ions typical of DNA. These observations are in agreement with the published dissociation mechanism of DNA, which assumes protonation of the nucleobase as the first step initiating subsequent base loss and backbone cleavage [3].

Our experiments revealed a striking difference between the product ion spectra of iron(III)-pentamer complexes and the spectra of uncomplexed oligonucleotides. Abundant peaks due to Fe(III)-complexed [M-T]-, w_4-, and [w_4-B]-fragment ions are observed. Additionally, the internal [w_4-d_4]-fragment ion is detected (Fig. 4). All fragment ions of high abundance are metal-

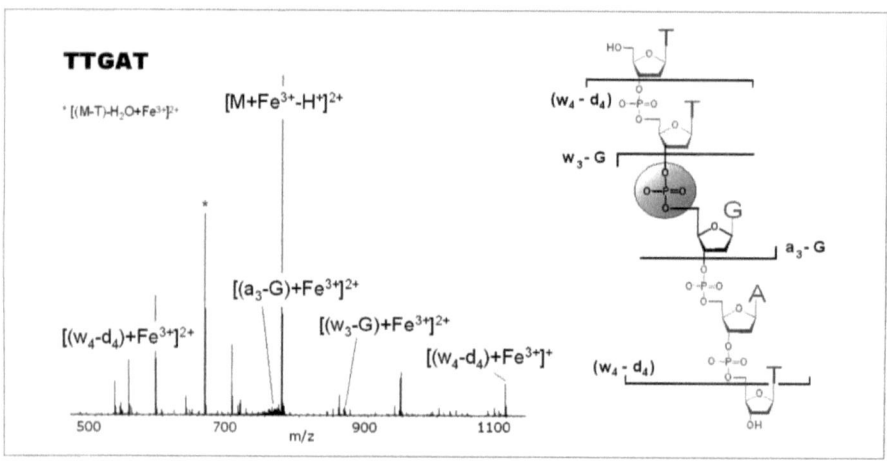

Fig. 4. Successful localization of the Fe(III) ion demonstrated for the pentadeoxynucleotide d(TTGAT). The [M+Fe^{3+}-H$^+$]$^{2+}$ ion was selected as the precursor ion. Key fragments providing information about the coordination site are indicated in the spectrum.

complexed, whereas small fragments, e.g. w_2, remain uncomplexed and are observed with very low abundance. These results strongly suggest a very tight bond between the Fe(III)-ion and the oligonucleotides. Furthermore, the Fe(III)-ion has a stabilizing effect on the central region of the oligonucleotide, as the $[w_4-d_4]$ core fragment is observed with high abundance. Apparently, it is not feasible to separate the Fe(III)-ion from the oligonucleotide under low-energy CID conditions.

Occurrence or absence of certain metal-complexed fragment ions provides information about the site of complexation [7]. Other authors suspected the nucleobases as the binding sites. However, considering the high abundance of metal-complexed fragments with cleaved nucleobase, like $[w_4-B]$, we cannot support this assumption. In our approach, localization of the binding site of the Fe(III)-ion was performed by identification of the shortest metal-complexed fragment ions from the 3'- and the 5'-end (Fig. 4). Simultaneous occurrence of the metal-complexed fragment ions $[w_3-B_3]$ and $[a_3-B_3]$ restricts the position of the Fe(III)-ion to the third nucleotide from the 3'-end. Since the non-terminal sugar residues do not provide any suitable binding site for metal ions, the Fe(III)-ion must coordinate to the third nucleotide. Coordination of the nucleobase can clearly be excluded, as both decisive fragment ions lack the nucleobase. Analyses of various Fe(III)-pentamer complexes gave identical results and the Fe(III)-ion can undoubtedly be assigned to the second phosphate group from the 5'-end (Fig. 4).

Extended studies on $[Fe(II)]_3$-pentamer and $[Zn(II)]_3$-pentamer complexes revealed very strong complexation as well, as no metal ions were released upon CID. Iron(III), iron(II) and zinc(II) ions all form extremely stable complexes with oligonucleotides and induce a strong stabilization of the phosphate backbone, thus resulting in greatly reduced fragmentation.

Tandem mass spectrometry provides insight into the complex world of metal-oligonucleotide complexes. The technique holds a great potential for characterization of naturally occurring complexes, localization of metal ions, and even investigation of metal ion induced conformational changes. Since the addition of metal ions affects the gas-phase behavior of oligonucleotides, it provides an alternative route for gaining sequence information.

Acknowledgments

We wish to thank the Swiss National Science Foundation for financial support of this work.

Received: August 17, 2005

[1] S.A. McLuckey, G.J. Van Berkel, G.L. Glish, *J. Am. Soc. Mass Spectrom.* **1992**, *3*, 60–70.
[2] E. Nordhoff, F. Kirpekar, P. Roepstorff, *Mass Spectrom. Rev.* **1996**, *15*, 67–138.
[3] Z. Wang, K.X. Wan, R. Ramanathan, J.S. Taylor, M.L. Gross, *J. Am. Soc. Mass Spectrom.* **1998**, *9*, 683–691.
[4] S. Schürch, E. Bernal-Méndez, C.J. Leumann, *J. Am. Soc. Mass Spectrom.* **2002**, *13*, 936–945.
[5] J.M. Tromp, S. Schürch, *J. Am. Soc. Mass Spectrom.* **2005**, *16*, 1262–1268.
[6] J.L. Beck, M.L. Colgrave, S.F. Ralph. M.M. Sheil, *Mass Spectrom. Rev.* **2001**, *20*, 61–87.
[7] S.T.M. Monn, S. Schürch, *J. Am. Soc. Mass Spectrom.* **2005**, *16*, 370–378.

Electrospray ionization tandem mass spectrometry of biphenyl-modified oligo(deoxy)ribonucleotides

Jan M. Tromp and Stefan Schürch

Rapid Communications in Mass Spectrometry
2006, 20, 2348-2354

In this work experiments with relatively long oligonucleotides incorporating biphenyl-modified building blocks were presented. The alcylic biphenyl nucleobase substitutes are C-glycosidic bound, which inhibit nucleobase loss and consequently influence the dissociation of oligonucleotides. DNA dissociation adjacent to the modified building blocks is not observed, which demonstrates the dependence of DNA backbone dissociation from base loss. In contrast to DNA dissociation, backbone fragmentation of RNA is not influenced by the biphenyl modification indicating the lack of dependence. In addition, experimental data show sequence verification of known RNA and DNA, provide information on alternative dissociation pathways and demonstrate the localization of a modified building block in a DNA sequence.

Electrospray ionization tandem mass spectrometry of biphenyl-modified oligo(deoxy)ribonucleotides

Jan M. Tromp and Stefan Schürch*
Department of Chemistry and Biochemistry, University of Bern, Bern, Switzerland
Received 3 April 2006; Revised 29 May 2006; Accepted 4 June 2006

Antisense oligonucleotides and aptamers are important candidates for future therapeutic applications. Different structural modifications are introduced into oligonucleotides to obtain high affinity and binding specificity. Sequence elucidation of oligonucleotides incorporating a wide variety of modifications presents an analytical challenge, as the standard protocols cannot be applied. Mass spectrometry has the potential to solve complex structural problems. However, a better understanding of the fundamental aspects of gas-phase dissociation of modified DNA and RNA is needed.

In this work the influence of specific chemical modifications on backbone dissociation is pointed out. Biphenyl-modified oligo(deoxy)ribonucleotides, which incorporate C-glycosidic bound abasic nucleobase substitutes, were subjected to collision-induced dissociation in an electrospray tandem mass spectrometer, with the goal to investigate the role of nucleobase loss on backbone dissociation. DNA bearing biphenyl nucleobase substitutes show abundant [a-B]- and w-ions generated by cleavage of the 3'-C–O bonds, except for the phosphodiester groups adjacent to the biphenyl modifications. At these positions no dissociation was observed, demonstrating the dependence of DNA backbone dissociation on nucleobase loss. Also, no evidence for a base loss independent mechanism responsible for formation of w-ions was found. RNA incorporating biphenyl nucleobase substitutes fragment into c- and y-ions resulting from cleavage of the 5'-P–O bond. Adjacent to the biphenyl modifications no altered dissociation behavior was found. This leads to the conclusion that dissociation of RNA is independent of the 1'-modification and, therefore, independent of nucleobase loss. Copyright © 2006 John Wiley & Sons, Ltd.

Oligonucleotide-based therapeutics offer a great potential for successful human cancer therapy. A variety of differently acting compounds is currently being evaluated for therapeutic applications[1–4] and great effort is put into the development of chemically modified oligonucleotides to direct the mechanism of action, to increase their affinity for target molecules, and to further enhance their binding specificity and biostability.[5–10] Apart from the development of appropriate synthetic methodologies, evaluation of nucleic-acid-based drug candidates by combinatorial strategies is among the main focuses of current research activities. Within this context, fast, reliable, and accurate tools for verification of the structural integrity and for detailed structural characterization of selected compounds are needed. The choice of techniques capable of characterizing highly modified oligonucleotides is considerably restricted. Considering the large diversity of structural elements potentially present in oligonucleotides, the overall size of the compounds and the small amount of material typically available for analysis, tandem mass spectrometry (MS/MS) is one of the most attractive analytical techniques for providing the high degree of structural information required. Structural modifications are identified due to the formation of structure-specific fragment ions generated upon collision-induced dissociation (CID). However, it has to be taken into consideration that most modifications lead to alteration of the dissociation behavior in the gas phase. Consequently, unambiguous structural elucidation by MS/MS along with the development of software-aided sequencing tools requires knowledge of the fundamental mechanistic aspects of the gas-phase dissociation of modified oligonucleotides.

Within the past decade successful sequencing of oligodeoxyribonucleotides of limited length by MS/MS has been demonstrated repeatedly. The mechanistic background of DNA dissociation in the gas phase has been investigated extensively and well-founded mechanistic data on the dissociation of unmodified oligodeoxyribonucleotides have been provided.[11–14] Dissociation of oligodeoxyribonucleotides is initiated by loss of the nucleobase, followed by cleavage of the 3'-C–O bond and formation of [a-B]- and w-ions.

*Correspondence to: S. Schürch, Department of Chemistry and Biochemistry, University of Bern, Freiestrasse 3, CH-3012 Bern, Switzerland.
E-mail: stefan.schuerch@ioc.unibe.ch
Contract/grant sponsor: University of Bern and the Swiss National Science Foundation.

Figure 1. Gas-phase dissociation of RNA is independent of nucleobase loss and primarily results in c- and y-ions.

The presence of the 2′-hydroxyl substituent in oligoribonucleotides (RNA) results in a significant change of the dissociation behavior, as it induces cleavage of the 5′-P–O bond and subsequent formation of c- and y-ions as the predominant dissociation products.[15–17] The central role the 2′-hydroxyl group plays in the dissociation of RNA, along with a dissociation mechanism, has been discussed previously.[18] The mechanism (Fig. 1) assumes backbone dissociation to be initiated by formation of an intramolecular cyclic transition state with the 2′-hydroxyl proton bridged to the 5′-phosphate oxygen. Subsequently, abstraction of the 2′-hydroxyl proton by the 5′-oxygen leads to dissociation of the 5′-P–O bond. After stabilization due to bond rearrangement, such mechanism results in release of complementary c- and y-type fragment ions. Investigation of RNA demonstrated that even small structural modifications, such as substitution of the 2′ hydrogen by a hydroxyl group, affect the gas-phase dissociation of oligonucleotides significantly, as they provide stabilization of the N-glycosidic bond or interfere with the main dissociation pathway directly.[19–23]

In the present work, we report on the dissociation of relatively long oligo(deoxy)ribonucleotides incorporating single or multiple biphenyl nucleobase substitutes. Oligonucleotides exhibit hydrophobic abasic sites bound by C-glycosidic bonds that inhibit dissociation via the known pathways. Precursor ions of relatively low charge states have been selected for CID to avoid excessive distribution of fragment ion current over multiple charge states and to minimize peak overlapping. Results indicate the dependence of backbone dissociation from base loss, point out the role of proton transfer reactions upon dissociation, and provide information on alternative pathways of fragment ion generation.

EXPERIMENTAL

Oligonucleotide synthesis and sample preparation

Biphenyl-modified 12- and 14-mer oligonucleotides were synthesized in-house on a 1 μmol scale using standard phosphoramidite chemistry.[24,25] Final concentrations were determined optically. For mass spectrometric experiments, oligonucleotides were dissolved in water, resulting in a final concentration of 25 pmol/μL, and sample volumes of 2 μL were used for analyses.

Mass spectrometry

All experiments were performed on an Applied Biosystems/MDS Sciex QStar Pulsar hybrid quadrupole-time-of-flight (QTOF) mass spectrometer (MDS Sciex, Concord, Canada), equipped with a nanoelectrospray ionization source (Proxeon Biosystems, Odense, Denmark). Oligonucleotides were analyzed in the negative-ion mode with a potential of −900 V applied to the nanospray needle. Nitrogen was used as the curtain gas.

Precursor ions for MS/MS experiments were selected within a window of ±1.5 m/z. CID was performed with collision energies in the range from 10 to 40 eV using nitrogen as the collision gas. The TOF analyzer was calibrated externally using a mixture of caesium iodide and taurocholic acid. Mass resolving power was in the range of 10 000 (FWHM). The Applied Biosystems Analyst QS 1.1 software package was used for data processing.

RESULTS AND DISCUSSION

Oligodeoxyribonucleotides

Figure 2 shows the product ion spectrum of the triply deprotonated 12-mer oligodeoxyribonucleotide d(GATGACXGCTAG), where X represents the biphenyl-substituted deoxyribose. Besides the presence of the precursor ion and the ion resulting from loss of a neutral guanidine (neutral nucleobase G), the spectrum is characterized by abundant, singly and doubly charged [a-B]- and w-ions generated by cleavage of the 3′-C–O bonds, which are the typical dissociation products of DNA. The series of [a-B]-ions from [a_2-A] through [a_9-C] and the complementary w-ions (w_1–w_8) are observed, thus providing extensive sequence information. However, the [a_7-X]-ion (m/z 986.67) and its complementary w_5-ion (m/z 790.12) are not observed (Table 1), indicating resistance of the C-glycosidic bond towards cleavage and complete suppression of backbone dissociation adjacent to the modification. Due to the abasic nature of the biphenyl modification and the presence of the C-glycosidic bond, protonation and base loss, which are the first steps of DNA backbone dissociation, are not observed.

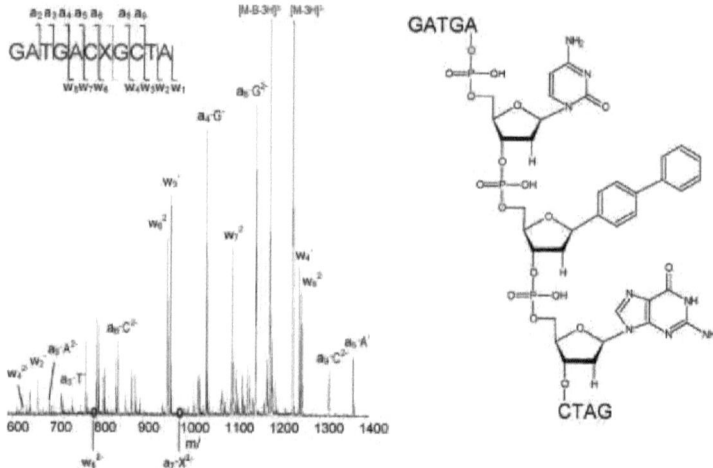

Figure 2. Product ion spectrum of the modified oligodeoxyribonucleotide d(GATGACXGCTAG), with X representing the 1′-biphenyl-modified building block (left). The spectrum is characterized by abundant [a-B]- and w-ions generated by cleavage of the 3′-C–O bond. No dissociation of the phosphodiester bond adjacent to the modified building block was observed. Structure of the 1′-biphenyl modification (right).

The corresponding gap in the fragment ion series indicates the position of the modified building block.

In the product ion spectra of oligodeoxyribonucleotides, w-ions are commonly of much higher abundance than their complementary [a-B]-ions. This observation led to the assumption that alternative fragmentation mechanisms exist which result in w-ions as well.[11,12,26] Since our experiments give no evidence of any fragmentation adjacent to the biphenyl-modified deoxyribose, the existence of any alternative fragmentation pathway that is independent of base loss can be excluded. If any alternative mechanism for the formation of w-ions does exist, it must be based on loss of the nucleobase as one of the initial reaction steps, and, consequently, is influenced by nucleobase modifications. Furthermore, dissociation of the phosphodiester backbone was found to be locally controlled, influenced by nearby functional groups only. The 1′-biphenyl modification alters dissociation of the adjacent phosphodiester group, while dissociation next to unmodified building blocks follows the DNA typical pathway.

Comparable results were obtained for the 14-mer oligodeoxyribonucleotide d(GATGACXXXGCTAG), which incorporates three consecutive central biphenyl building blocks. Figure 3 shows the corresponding product ion spectrum generated by dissociation of the [M−4H]$^{4-}$ precursor ion with m/z 1092.96. Dissociation of the oligonucleotide into [a-B]- and w-ions was observed at the phosphodiester groups adjacent to unmodified nucleotides, thus providing sequence information. Lack of fragment ions (w$_5$, w$_6$, w$_7$, and [a$_7$-X], [a$_8$-X], [a$_9$-X]) due to the uncleavable C-glycosidic bond indicates the presence of modified building blocks. Relative abundances of the fragment ions of biphenyl-modified dodeca- and tetradecanucleotides are summarized in Table 1. Besides the peaks referring to backbone cleavage, the spectrum shows additional abundant ions due to loss of nucleobases. Loss of negatively charged nucleobases is known to be a major fragmentation pathway for highly charged oligonucleotides.[27] However, in the case of triply and quadruply charged 12- and 14-mers, neutral base loss was found to be favored over anionic base loss. The preference of neutral base loss followed the order of G > A > C, resulting in ions with m/z 1055.19, 1059.19, and 1065.20, respectively. An ion with m/z 1061.45, indicating loss of an uncharged T, was not observed. The observed order of neutral base loss is in agreement with the order of the proton affinities of the nucleobases.[21,27,28]

The results are in agreement with the generally accepted mechanism for backbone dissociation of DNA, which is initiated by protonation of the nucleobase followed by anionic or neutral base loss.[12] As neither protonation of the biphenyl modification nor cleavage of the C-glycosidic bond can take place, DNA typical dissociation via the known pathway is blocked. Experiments clearly demonstrate that DNA dissociation is dependent on nucleobase loss. Also, there is no evidence for any alternative mechanism responsible for formation of w-ions.

Oligoribonucleotides

Oligoribonucleotides (RNA) bearing a single and three biphenyl nucleobase substitutes, respectively, were subjected to CID to study the dissociation mechanism of RNA and its dependence on base loss. Dissociation of this class of

ESI-MS/MS of biphenyl oligo(deoxy)ribonucleotides

Table 1. Comparison of the fragment ion distribution of four biphenyl-modified oligo(deoxy)ribonucleotides

	d(GATGACXGCTAG) M = 3711.69 Da [M−3H]$^{3-}$									(GAUGACXGCUAG) M = 3875.60 Da [M−3H]$^{3-}$								
Base	a_n/a_n-B	b_n	c_n	d_n	w_n	x_n	y_n	z_n	n	Base	a_n/a_n-B	b_n	c_n	d_n	w_n	x_n	y_n	z_n	n
G	0	0	2	0	0	0	0	0	1	G	0	0	25	0	0	0	16	0	11
A	81	1	4	0	7	0	0	0	2	A	28	0	74	0	0	0	4	0	10
T	10	2	2	0	0	0	0	0	3	U	13	0	43	0	0	0	8	2	9
G	100	0	0	0	31	0	0	0	4	G	19	0	9	2	7	0	1	3	8
A	26	0	0	0	41	0	0	0	5	A	5	0	11	3	8	0	15	0	7
C	32	0	0	0	66	0	0	0	6	C	8	0	28	0	11	3	10	3	6
X	0	0	0	0	0	0	0	0	7	X	0	0	25	0	0	0	7	0	5
G	65	0	0	0	54	0	0	0	8	G	0	0	11	0	10	0	12	0	4
C	11	0	0	0	78	2	2	0	9	C	0	0	8	0	40	0	50	0	3
T	0	0	0	0	14	4	1	0	10	U	0	0	4	0	14	0	100	0	2
A	7	0	0	0	65	2	0	0	11	A	0	0	20	0	22	0	0	0	1
G										G									

	d(GATGACXXXGCTAG) M = 4375.86 Da [M−4H]$^{4-}$									(GAUGACXXXGCUAG) M = 4571.75 Da [M−4H]$^{4-}$								
Base	a_n/a_n-B	b_n	c_n	d_n	w_n	x_n	y_n	z_n	n	Base	a_n/a_n-B	b_n	c_n	d_n	w_n	x_n	y_n	z_n	n
G	0	0	1	0	0	0	0	0	1	G	0	0	38	0	0	0	8	0	13
A	47	0	0	0	6	0	0	0	2	A	34	0	67	0	0	0	7	0	12
T	8	3	0	0	0	0	0	0	3	U	25	0	62	0	0	0	14	0	11
G	35	0	0	0	*	0	0	0	4	G	15	0	31	11	0	0	15	0	10
A	100	0	0	0	64	0	0	0	5	A	32	0	39	7	0	0	5	0	9
C	46	0	0	0	24	0	0	0	6	C	18	0	41	0	7	0	11	0	8
X										X									

Continues

Table 1. Continued

	d(GATGACXXXGCUAG) M = 4375.86 Da [M-4H]$^{4-}$								Base	n	(GAUGACXXXGCUAG) M = 4571.75 Da [M-4H]$^{4-}$							n
Base	n	a_n/a_n-B	b_n	c_n	d_n	w_n	x_n	y_n	z_n			a_n/a_n-B	b_n	c_n	d_n	w_n	x_n	y_n	z_n
X	7	0	0	0	0	0	0	0	0	X	7	0	0	19	0	0	0	21	0
X	8	0	0	0	0	0	0	0	0	X	8	0	0	10	0	14	0	29	0
G	9	33	0	0	0	91	0	0	0	G	9	0	0	5	0	0	0	43	7
C	10	0	0	0	0	42	0	3	0	C	10	0	0	8	0	39	0	13	0
T	11	7	0	0	0	13	0	0	0	U	11	0	0	11	0	38	0	57	0
A	12	0	0	0	0	33	1	0	0	A	12	0	0	8	0	17	0	100	0
G	13	6	0	0	0	0	0	0	0	G	13	0	0	0	0	25	0	0	0

Most abundant DNA fragments are [a-B]- and w-ions, whereas no fragmentation adjacent to the biphenyl modification was observed. RNA typical fragmentation results in c- and y-ions. No influence of the modification on the fragmentation behavior could be found. Numbers indicate relative peak intensities with the highest peak normalized to 100. Intensities of a_n- and $[a_n-B]$-ions are combined.
*The abundance of w_{10}^{3-} of d(GATGACXXXGCUAG) could not be determined, since the ion overlaps with [M-HA]$^{1+}$.

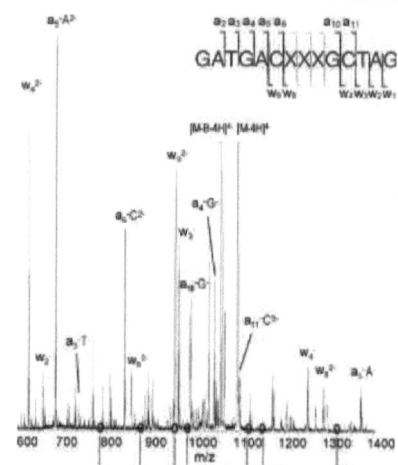

Figure 3. Product ion spectrum of d(GATGACXXXGCTAG), showing DNA typical fragmentation. Lack of backbone cleavage adjacent to the biphenyl-modified building blocks points out the dependence of DNA dissociation on nucleobase loss.

compounds has hardly been explored by MS/MS. A dissociation mechanism of RNA has been recently proposed, supported by experiments on specifically modified tetra-ribonucleotides.[18] Data demonstrated the key role of the 2'-hydroxyl proton in dissociation of RNA.

Base loss from RNA is observed, without stringently resulting in backbone cleavage. Subjecting dodecaribonucleotides with C-glycosidic bound substituents to CID will provide more conclusive data about the influence of base loss on the gas-phase dissociation of RNA. The product ion spectrum of GAUGACXGCUAG ([M–3H]$^{3-}$), an all-2'-hydroxyl dodecamer with a 1'-biphenyl-modified building block, is shown in Fig. 4. The oligoribonucleotide primarily dissociates into singly charged ions due to cleavage of the 5'-P-O bond. The two oppositely directed RNA typical fragment ion series (c_1–c_{11} and y_2–y_{11}) provide complete sequence information for the 12-mer. The y_1 ion was not detected, since it does not provide any phosphate group as a deprotonation site. In contrast to the dissociation of the modified DNA sequence, which did not show any backbone cleavage at the position of the modification, dissociation of the RNA-based oligonucleotide was independent of the modification. The phosphodiester group adjacent to the modification shows identical fragmentation behavior as the unmodified sections of the sequence. RNA typical cleavage of the 5'-P-O bond was observed, resulting in a c_7-ion with m/z 1152.66, and singly and doubly charged y_5-ions with m/z 1567.25 and 783.12, respectively. There is no evidence for any influence of the 1'-biphenyl modification on backbone dissociation, as shown in Table 1.

Even the presence of three consecutively arranged biphenyl modifications within a 14-mer oligoribonucleotide (GAUGACXXXGCUAG) does not alter the fragmentation

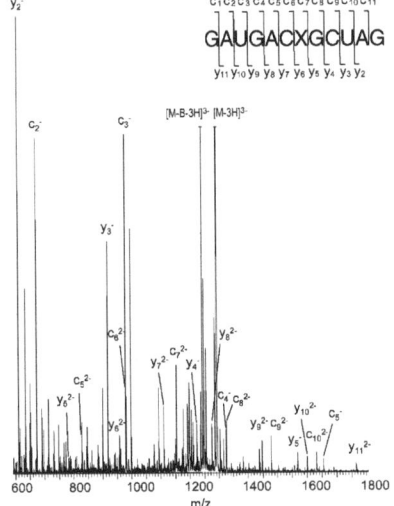

Figure 4. Product ion spectrum of GAUGACXGCUAG predominantly shows RNA typical c- and y-ions and provides complete sequence information.

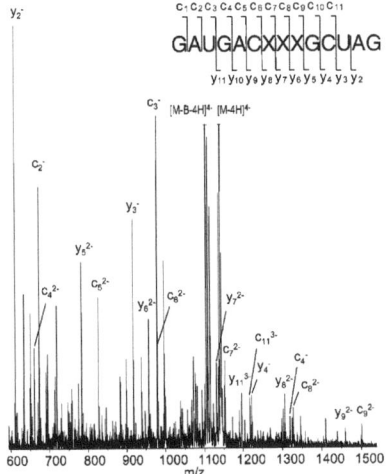

Figure 5. The product ion spectrum of GAUGACXXXG-CUAG demonstrates that there is no influence of the biphenyl modification on RNA backbone dissociation.

behavior. The complete nucleotide sequence can be identified unambiguously from the product ion spectrum of the 14-mer, as demonstrated in Fig. 5. The quadruply charged precursor ion selected for CID results in fragment ions of higher charge states as well. Due to the high mass resolving power achieved upon analysis, peak overlapping is less prominent and peak assignment is still unambiguous. Backbone cleavage within the central part of the sequence bearing three consecutive biphenyl modifications occurs in a RNA typical fashion with c- and their complementary y-ions formed as the predominant dissociation products. Additionally, short w-ions are formed with noticeable abundance.

These findings emphasize the independence of RNA dissociation from the nucleobase and support the previously proposed dissociation mechanism, as shown in Fig. 1. The centrally located phosphodiester groups of GAUGACXXXG-CUAG are cleaved in the same manner, regardless of the lack of any directly adjoining nucleobase. Thus, the influence of an adjacent nucleobase or any remotely located functional group on backbone dissociation of RNA can be excluded.

CONCLUSIONS

Tandem mass spectrometric investigation of oligodeoxyribonucleotides and oligoribonucleotides bearing a single or multiple 1'-biphenyl-modified building blocks gives evidence for the influence of the nucleobase on the gas-phase dissociation of this important class of compounds. DNA dissociation is characterized by cleavage of the 3'-C-O bonds resulting in complete series of [a-Base]- and w-ions, whereas cleavage of the backbone adjacent to the biphenyl-modified building blocks does not take place, demonstrating unambiguously the dependence of DNA backbone dissociation on nucleobase loss. The C-glycosidic bond of the modified building block inhibits backbone cleavage through the known dissociation channels and there is no evidence for any alternative fragmentation pathway.

In contrast, the preferred fragmentation sites of RNA are the 5'-P-O bonds, leading to c- and y-ions. No difference in gas-phase dissociation was found between unmodified nucleotides and biphenyl-modified building blocks. Backbone dissociation adjacent to the biphenyl-modified building blocks results in c- and y-ions as well. These results demonstrate that, unlike DNA, gas-phase dissociation of RNA is independent of nucleobase loss, at least for the range of relatively low charge states examined in the present work.

In-depth investigation of the mechanistic aspects of oligonucleotide dissociation is needed for development of MS-based sequencing methods to be applied to modified oligonucleotides. Experiments point out the differences between DNA and RNA dissociation in the gas phase and clarify the influence of the nucleobase on these mechanisms.

Acknowledgements

We thank Dr. Christine Brotschi and Prof. Christian Leumann for generously providing synthetic oligonucleotides and the University of Bern and the Swiss National Science Foundation for generous financial support of this work.

REFERENCES

1. Leumann CJ. *Bioorg. Med. Chem.* 2002; **10**: 841.
2. Herbst RS, Frankel SR. *Clin. Cancer Res.* 2004; **10**: 4245.
3. Yacyshyn BR, Chey WY, Goff J, Salzberg B, Baerg R, Buchman AL, Tami J, Yu R, Gibiansky E, Shanahan WR. *Gut.* 2002; **51**: 30.
4. Sewell KL, Geary RS, Baker BF, Glover JM, Mant TG, Yu RZ, Tami JA, Dorr A. *J. Pharmacol. Exp. Ther.* 2002; **303**: 1334.
5. Tang W, Zhu L, Smith LM. *Anal. Chem.* 1997; **69**: 302.
6. Baker BF, Monia BP. *Biochim. Biophys. Acta* 1999; **1489**: 3.
7. Goodchild J. *Curr. Opin. Mol. Ther.* 2004; **6**: 120.
8. Altmann K-H, Dean NM, Fabbro D, Freier SM, Geiger T, Häner R, Hüsken D, Martin P, Monia BP, Müller M, Natt F, Nicklin P, Phillips J, Piels U, Sasmor H, Moser HE. *Chimia* 1996; **50**: 168.
9. Myers KJ, Dean NM. *TiPS* 2000; **21**: 19.
10. Beck JL, Colgrave ML, Ralph SF, Sheil MM. *Mass Spectrom. Rev.* 2001; **20**: 61.
11. Wan KX, Gross J, Hillenkamp F, Gross ML. *J. Am. Soc. Mass Spectrom.* 2001; **12**: 193.
12. Wang Z, Wan KX, Ramanathan R, Taylor JS, Gross ML. *J. Am. Soc. Mass Spectrom.* 1998; **9**: 683.
13. McLuckey SA, Habibi-Goudarzi S. *J. Am. Chem. Soc.* 1993; **115**: 12085.
14. McLuckey SA, Van Berkel GJ, Glish GL. *J. Am. Soc. Mass Spectrom.* 1992; **3**: 60.
15. Cerny RL, Tomer KB, Gross ML, Grotjahn L. *Anal. Biochem.* 1987; **165**: 175.
16. Kirpekar F, Krogh TN. *Rapid Commun. Mass Spectrom.* 2001; **15**: 8.
17. Schürch S, Bernal-Mendez E, Leumann CJ. *J. Am. Soc. Mass Spectrom.* 2002; **13**: 936.
18. Tromp JM, Schürch S. *J. Am. Soc. Mass Spectrom.* 2005; **16**: 1262.
19. Nordhoff E, Cramer R, Karas M, Hillenkamp F, Kirpekar F, Kristiansen K, Roepstorff P. *Nucleic Acids Res.* 1993; **21**: 3347.
20. Bartlett MG, McCloskey JA, Manalili S, Griffey RH. *J. Mass Spectrom.* 1996; **31**: 1277.
21. Wan KX, Gross ML. *J. Am. Soc. Mass Spectrom.* 2001; **12**: 580.
22. Wang BH, Hopkins CE, Belenky AB, Cohen AS. *Int. J. Mass Spectrom. Ion Processes* 1997; **169/170**: 331.
23. Sannes-Lowery KA, Hofstadler SA. *J. Am. Soc. Mass Spectrom.* 2003; **14**: 825.
24. Brotschi C, Mathis G, Leumann CJ. *Chem. Eur. J.* 2005; **11**: 1911.
25. Brotschi C, Leumann CJ. *Chem. Commun.* 2005; **15**: 2023.
26. Pomerantz SC, McCloskey JA. *Anal. Chem.* 2005; **77**: 4687.
27. Pan S, Verhoeven K, Lee JK. *J. Am. Soc. Mass Spectrom.* 2005; **16**: 1853.
28. Greco F, Liguori A, Sindona G, Uccella N. *J. Am. Chem. Soc.* 1990; **112**: 9092.

Mass Spectrometry of Oligonucleotides

Stefan Schürch, Jan M. Tromp and Selina T. M. Monn

Nucleosides, Nucleotides and Nucleic Acids
2007, 26, 1629-1633

This article gives detailed overview about gas-phase dissociation of oligonucleotides. It describes the different fragmentation products and explains the corresponding dissociation mechanisms. Based on the knowledge of dissociation behavior, an oligoribonucleotide is sequenced. The RNA typical c- and y-ions are generated by cleavage of the 5'-P-O bond and provide full sequence coverage. In addition, experiments studying the influence of various 1'- and 2'-substituents on backbone cleavage are presented.

MASS SPECTROMETRY OF OLIGONUCLEOTIDES

Stefan Schürch, Jan M. Tromp, and Selina T. M. Monn ◻ *Department of Chemistry and Biochemistry, University of Bern, Bern, Switzerland*

◻ *Expanding research in the field of modified oligonucleotides demands suitable analytical tools for size and purity verification of known compounds and accurate structure elucidation of unknowns. There is a need for characterization of the types and sites of modifications in oligonucleotides and to identify and sequence selected candidates originating from synthesis. The potential of electrospray tandem mass spectrometry (ESI-MS/MS) for structural characterization and sequencing of oligonucleotides is demonstrated. The fundamental behavior of DNA, RNA, and selected modified oligonucleotides in gas-phase is shown. Since gas-phase dissociation does not demand specific structural prerequisites, the method bears a great potential for rapid and most accurate characterization of modified oligonucleotides, e.g. from combinatorial libraries.*

Keywords Oligonucleotide sequencing; gas-phase dissociation; electrospray; tandem mass spectrometry

INTRODUCTION

Oligonucleotide-based therapeutics bear a great potential for successful human cancer therapy. A variety of compounds is evaluated for therapeutic applications and great effort is put into the development of chemically modified oligonucleotides with the goal to direct the mechanism of action, to increase their affinity towards target sequences, and to enhance their binding specificity and biostability. Apart from the development of appropriate synthetic methodologies, evaluation of nucleic acid-based drug candidates is among the main focuses of current research activities. Within this context, fast, reliable, and accurate tools are needed for verification of the structural integrity and for detailed structural characterization of selected compounds.

Tandem mass spectrometry (MS/MS) is a powerful tool for rapid structure confirmation and elucidation. The technique is based on selection of precursor ions in a first stage of mass spectrometry, followed by activation of the precursor ions by collisions with an inert gas. Collision-induced dissociation (CID) results in sequence- or structure-defining fragment ions, which

Address correspondence to Stefan Schürch, Department of Chemistry and Biochemistry, University of Bern, Freiestrasse 3, CH-3012 Bern, Switzerland. E-mail: stefan.schuerch@ioc.unibe.ch

are finally separated and detected in a second stage of mass spectrometry. In combination with soft ionization techniques, such as electrospray ionization (ESI), tandem mass spectrometry has demonstrated its potential for routine peptide sequencing and protein identification. Despite its many advantages, tandem mass spectrometry is hardly employed for sequence elucidation of oligonucleotides. This is mostly due to the fact that the resulting product ion spectra are very complex and that the fundamental aspects of oligonucleotide dissociation in gas-phase are not understood completely.

SEQUENCING OF UNMODIFIED DNA AND RNA BY TANDEM MASS SPECTROMETRY

Unlike chemical or enzymatic degradation, gas-phase dissociation of oligonucleotides in the collision cell of a tandem mass spectrometer does not require specific structural prerequisites. The nucleotide sequence is read from the product ion spectrum by assigning one or more series of fragment ions generated by cleavage of the phosphodiester bond at various positions along the backbone. The main mechanism of backbone cleavage of DNA in gas-phase was solved by Gross and co-workers almost a decade ago.[1] It was demonstrated that backbone cleavage is initiated by protonation and subsequent loss of the nucleobase. Bond rearrangement finally results in cleavage of the 3'-C-O bond, generating the [a-Base]- and w-ions as the main dissociation products. Consequently, sequence elucidation of DNA is primarily based on assignment of these two ion series. The nomenclature of oligonucleotide fragment ions shown in Figure 1, is based on the initial work of McLuckey et al.[2]

The fragment ion pattern generated by dissociation of RNA clearly differs from the one of DNA. RNA predominantly dissociates by cleavage of the 5'-P-O bond, resulting in c- and their complementary y-ions as the main products.[3-5] The mechanism responsible for formation of the RNA-typical fragment ions has recently been elucidated by Tromp et al.[6,7] by studying the influence of various 1'- and 2'-substituents on backbone cleavage. It was demonstrated that, in contrast to the dissociation of DNA, the nucleobase adjacent to the cleavage site is not involved in the cleavage mechanism of RNA. The mechanism assumes that backbone cleavage is initiated by formation of an intramolecular cyclic transition state with the 2'-hydroxyl proton bridged to the 5'-oxygen. Abstraction of the 2'-hydroxyl proton by the 5'-phosphate oxygen leads to scission of the 5'-P-O bond and finally, results in the release of a y-fragment with intact 5'-terminus and a stabilized negatively charged c-fragment ion. The independence of RNA backbone cleavage from nucleobase loss has further been demonstrated by experiments on 1'-biphenyl-substituted oligonucleotides.[8] Furthermore, experiments on chimeric oligonucleotides containing deoxyribo- and ribonucleotides

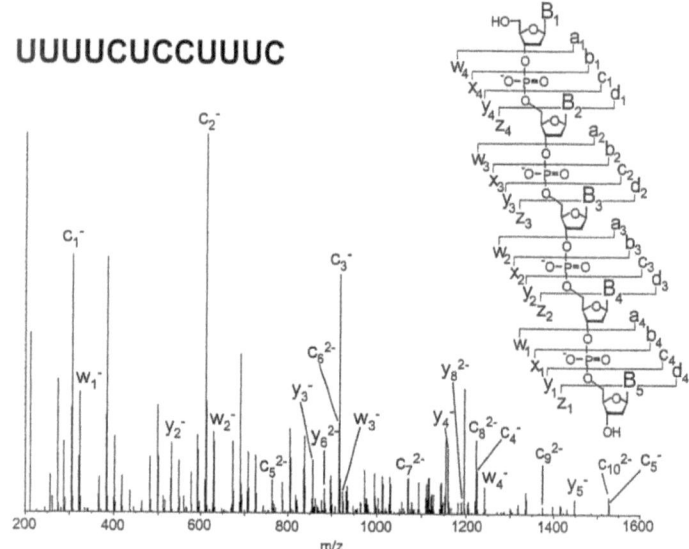

FIGURE 1 Product ion spectrum of the dodecaribonucleotide UUUUCUCCUUUC (3606 Da). Sequence-defining c- and y-ions are generated by cleavage of the 5'-P-O bond and provide full sequence coverage. Right: General nomenclature of ions generated by gas-phase dissociation of oligonucleotides.

demonstrated that backbone cleavage is locally controlled, influenced by the adjacent functional groups only.[5] An example of RNA sequencing by tandem mass spectrometry is given by Figure 1, which shows the product ion spectrum of the dodecaribonucleotide UUUUCUCCUUUC (3606 Da). The triply charged molecular ion with m/z 1201.1 was selected as the precursor ion and subjected to collision-induced dissociation. The spectrum shows the two series of oppositely directed c- and y-type fragment ions as the main dissociation products. Additional sequence information is obtained from the w-ion series. Besides these main series of sequence-defining fragment ions, numerous fragments generated by alternative backbone cleavage are observed as well, though with less abundance.

SEQUENCING OF MODIFIED OLIGONUCLEOTIDES

Due to the unnatural structural elements present in modified oligonucleotides, they often resist degradation by conventional methods. Since gasphase dissociation does not depend on the activity of a specific cleavage

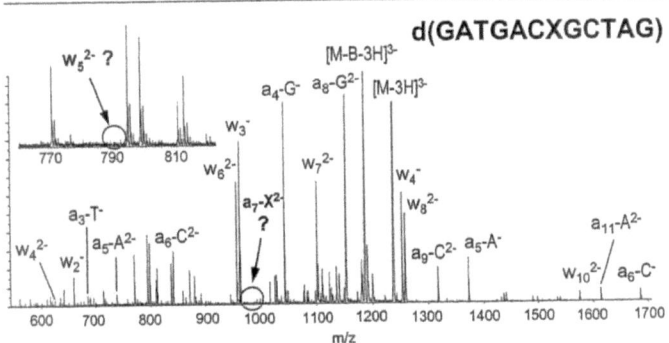

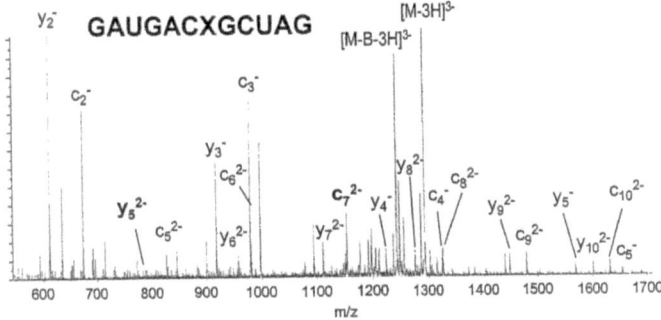

FIGURE 2 Product ion spectra of the 1'-biphenyl-modified dodecadeoxyribonucleotide d(GATGACXGCTAG) and the corresponding dodecaribonucleotide GAUGACXGCUAG, with X indicating the 1'-biphenyl modification.

reagent, tandem mass spectrometry is a promising alternative for characterization of such compounds. Product ion spectra offer rapid read-out of nucleotide sequences and simultaneously, provide information on types and positions of modifications. Product ion spectra of oligonucleotides incorporating a 1'-biphenyl modification are shown in Figure 2. Since backbone dissociation of DNA depends on loss of the nucleobase, the series of [a-B]- and w-ions show a gap at the position of the modification. In the spectrum of d(GATGACXGCTAG) the w_5^{2-} and the a_7-X^{2-} fragment ion with m/z 790.12 and m/z 986.67, respectively, are not observed. On the other hand, RNA dissociation is independent of nucleobase loss and the product ion spectrum of GAUGACXGCUAG shows the uninterrupted series of sequence-defining c- and y-fragment ions, including the c_7^{2-} (m/z 1152.66) and the y_5^{2-} (m/z 783.12) ion.

CONCLUSIONS AND OUTLOOK

Structural characterization and sequence determination of oligonucleotides by ESI-MS/MS is an accurate and sensitive technique. Isolation of precursor ions is highly selective, thus, allowing mixture analysis. Dissociation of oligonucleotides in the collision cell of the mass spectrometer does not require specific cleavage reagents and is tolerant towards structural modifications. The high sensitivity of the method and the high degree of structural information obtained make the method highly attractive for sequence determination of modified oligonucleotides and oligonucleotides from combinatorial libraries.

In contrast to other biopolymers, oligonucleotides generate a large number of fragment ions upon CID. The presence of alternative dissociation products, whose mechanisms of formation are not known yet, results in highly complex spectra. Overlapping isotopic peaks patterns of fragments exhibiting similar m/z render peak assignment and data interpretation difficult. Further refinement of sequencing protocols are sought to direct backbone dissociation and generate simplified fragment ion patterns which enable fast and unambiguous read-out of the nucleotide sequences.

REFERENCES

1. Wang, Z.; Wan, K.X.; Ramanathan, R.; Taylor, J.S.; Gross, M.L. Structure and fragmentation mechanism of isomeric T-rich oligonucleotides: A comparison of four tandem mass spectrometric methods. *J. Am. Soc. Mass Spectrom.* **1998**, 9, 683–691.
2. McLuckey, S.A.; Van Berkel, G.J.; Glish, G.L. Tandem mass spectrometry of small, multiply charged oligonucleotides. *J. Am. Soc. Mass Spectrom.* **1992**, 3, 60–70.
3. Cerny, R.L.; Tomer, K.B.; Gross, M.L.; Grotjahn, L. Fast atom bombardment combined with tandem mass spectrometry for determining structures of small oligonucleotides. *Anal. Biochem.* **1987**, 165, 175–182.
4. Kirpekar, F.; Krogh, T.N. RNA fragmentation studied in a matrix-assisted laser desorption/ionization tandem quadrupole/orthogonal time-of-flight mass spectrometer. *Rapid Commun. Mass Spectrom.* **2001**, 5, 8–14.
5. Schürch, S.; Bernal-Méndez, E.; Leumann, C.J. Electrospray tandem mass spectrometry of mixed-sequence RNA/DNA oligonucleotides. *J. Am. Soc. Mass Spectrom.* **2002**, 3, 936–945.
6. Tromp, J.M.; Schürch, S. Gas-phase dissociation of oligoribonucleotides and their analogs studied by electrospray ionization tandem mass spectrometry. *J. Am. Soc. Mass Spectrom.* **2005**, 16, 1262–1268.
7. Monn, S.T.M.; Tromp, J.M.; Schürch, S. Mass spectrometry of oligonucleotides. *Chimia*, **2005**, 59, 822–825.
8. Tromp, J.M.; Schürch, S. Electrospray ionization tandem mass spectrometry of biphenyl-modified oligo(deoxy)ribonucleotides. *Rapid Commun. Mass Spectrom.* **2006**, 20, 2348–2354.

III. Discussion

Discussion

The results and discussion part in the previous publications is the basis of the discussion written here below. Therefore, all spectra and schemes, as well as descriptions of different fragmentation mechanisms, are found there. However, the discussion is continued and extended in the following part, in order to elucidate additional aspects of the experimental results, allowing a more detailed vision and a deepened understanding of tandem mass spectrometric oligonucleotide dissociation.

Dissociation products

Detailed structural analysis of the oligonucleotide was performed with tandem mass spectrometry. MS/MS employs collision induced dissociation (CID) to fragment a precursor ion. Low collision energies were used for the dissociation of the weakest bonds only. However, fragment ions generated by backbone cleavage at most positions were found in all experiments. But there are distinctive differences in abundances of the various types of fragment ions.

DNA

The primary product ions generated by CID of oligodeoxyribonucleotides (DNA) are [a-Base]- and w-ions. These fragment ions result from the backbone cleavage of the 3'-C-O bond. The DNA dissociation mechanism is initiated by the nucleobase loss. This is the reason why no a-ions appear but only [a-B]-ions. The complementary w-ions are in the majority of cases the most intensive fragment ions in product ion spectra.

The high abundance of the w-ions allows fast sequence verification of a known oligodeoxyribonucleotide. Why the w-ions are often clearly more abundant than their complementary [a-B]-ions is not yet completely understood. Possibly there is a better ionization of the 3'-terminal w-ions or there is an alternative fragmentation mechanism. Experiments with DNA which have a 1'-biphenyl modification show that there are no w-ions generated if the dissociation is not induced by the base loss. These experiments exclude the existence of any w-ion generating alternative dissociation mechanisms which are independent of nucleobase loss. Therefore, an alternative fragmentation pathway for the formation of w-ions must be based on nucleobase loss as an initiation step of the DNA dissociation mechanism.

RNA

The fragmentation pattern of oligoribonucleotides (RNA) looks even more complex than the one of DNA. Most of the possible fragment ions originating from backbone cleavage are found in the product ion spectra. The preferred dissociation products of RNA dissociation are the c- and their complementary y-ions. They are generated by the backbone cleavage of the 5'-P-O bond. In contrast to the w-ions in DNA spectra, RNA typical c- and y-ions are not so all-dominant. Nevertheless, c- and y-ions are useful for unambiguous sequence verification of a known oligoribonucleotide.

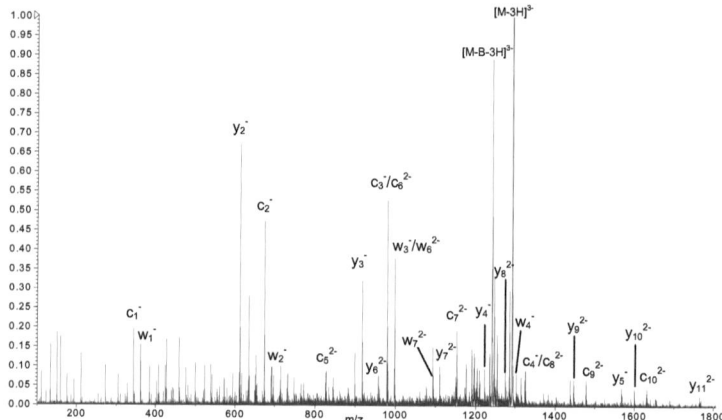

Figure 7. The product ion spectrum of RNA is characterized by abundant c- and y-ions, which provide complete sequence verification. However, also w-ions were observed.

Figure 7. shows a product ion spectrum of a RNA 12-mer. The complete c- and y-ion series can be identified non-stop, providing sequence verification. The y_1 ion can never be detected as there is no deprotonation site. The intensity of complementary c- and y-ions of the same size is comparable. No evidence of a preferred 3'- or 5'-terminal ionization is found. However, the relatively high abundance of w-ions in RNA spectra supports the idea of an alternative w-ions generating dissociation mechanism.

Local control

For studying details of the gas-phase dissociation mechanism oligonucleotides with a single modification were introduced. Modification of only one functional group or one nucleobase allows changes in the fragmentation pattern to be attributed directly to the modification. Very important for the interpretation of modified oligonucleotide spectra is the fact that most modifications have only a local influence. Experiments demonstrate that dissociation of the phosphodiester backbone is locally controlled and in that way only affected by its direct proximity. The fragmentation behavior of the phosphodiester group next to the modification is altered, but phosphodiester groups further away are not influenced. They show a fragmentation behavior typical of their own chemical vicinity.

Tetramer

The first experiments were performed with tetramers. Short oligonucleotides were chosen because this choice makes the interpretation of spectra easier and they can be dissociated very efficiently under CID conditions. In addition, tetramers are still large enough to allow the analysis of the influence of a selected modification on the backbone dissociation without the effect of terminal fragmentation, which is reflected sometimes in a slightly different fragmentation pattern.

For an example of the local control of backbone dissociation, a product ion spectrum of a RNA tetramer with a single 2'-fluoro modification at the second building block is presented in Figure 8.

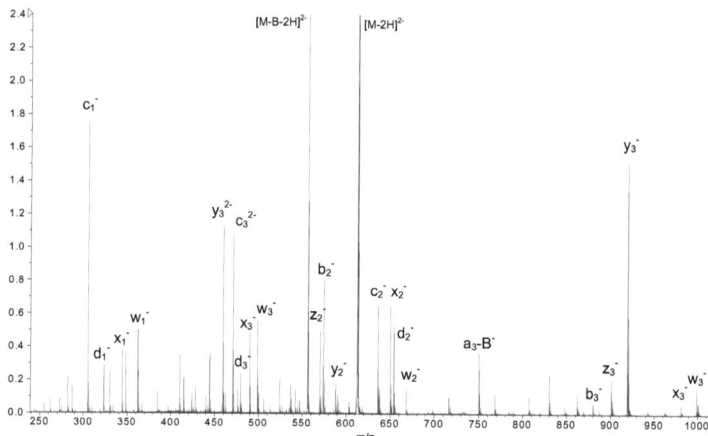

Figure 8. Product ion spectrum of a RNA tetramer incorporating a 2'-fluoro modification demonstrates that the modification alters locally the fragmentation behavior of the adjacent second phosphodiester group.

The spectrum shows for the first and third phosphodiester groups a fragmentation pattern typical of RNA. The complementary c_1/y_3 ions for the first and the c_3 ion for the third phosphodiester group are the predominant fragmentation products. y_1 can not be observed because of the lack of a protonation site. The second phosphodiester group, which is in the proximity of the 2'-fluoro modification, shows altered fragmentation behavior. No evidence of an influence of the modification on the rest of the oligonucleotide could be found.

Modifying the functional group at the 2' position of the ribose directly changes the backbone dissociation of the adjacent phosphodiester group. Mixed-sequence DNA/RNA oligonucleotides, for example, show different fragmentation patterns for every phosphodiester group, corresponding to their adjacent ribose.

Biphenyl oligodeoxyribonucleotides

Experiments with biphenyl-modified DNA demonstrate the local influence a substituted nucleobase can have on the backbone dissociation. Figure 9. shows the product ion spectrum of a DNA 12-mer. Intensive DNA typical [a-B]- and w-ions mold the spectrum. But at the position of the modification no fragmentation products are observed.

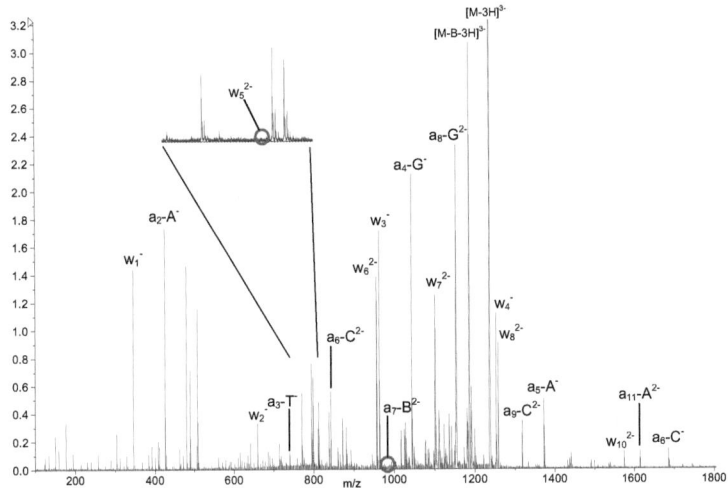

Figure 9. Product ion spectrum of DNA incorporating a single 1'-biphyl-modified building block. Highlighted are m/z values of w_5^{2-} and a_7-B^{2-}, which would be generated by the backbone cleavage adjacent to the modification.

The modification of the base results in the lack of backbone cleavage of the adjacent phosphodiester group, more precisely of the 3' phosphodiester group of the modified nucleoside. This is not obvious because the 5' phosphodiester group is sterically nearer to the base. But it can be understood according to the DNA dissociation mechanism of

Wang (Figure 2.) in which the nucleobase loss initiates the cleavage of the 3' phosphodiester group.

Conclusions

A good understanding of the local influence of the modification on the backbone dissociation allows the direct comparison of natural building blocks with modified ones. This is fundamental for the elucidation of the gas-phase dissociation mechanism, as it provides the opportunity to analyze the influence of every functional group participating in the RNA dissociation mechanism.

2'-modifications

Gas-phase dissociation of DNA and RNA differ conspicuously in terms of the product ions generated by CID. DNA dissociate dominantly in [a-B]- and w-ions, RNA in c- and y-ions. The only difference is the 2'-position of the ribose. In the case of DNA the ribose bears only a hydrogen; in RNA it bears hydroxyl group. This simple modification results in totally different fragmentation patterns, which led to the assumption that the backbone dissociation of DNA and RNA is based on two different mechanisms.

Nucleobase loss is known as the initiation step of DNA backbone fragmentation. In RNA spectra, fragment ions with nucleobase loss are hardly ever seen. This indicates that the 2'-hydroxyl group of RNA hinders the loss of a nucleobase. To analyze the influence of the 2'-functional group on the nucleobase loss and consequently also on the backbone dissociation several different 2'-modified oligonucleotides were investigated.

2'-methoxy

The methoxy modification was chosen because it bears an oxygen atom on the 2'-position of the ribose. This allows a direct comparison with RNA. The backbone fragmentation of the 2'-methoxy modified oligonucleotide shows no preferred fragmentation products. All possible fragment ions generated by cleavage of the phosphodiester backbone were found. As in RNA dissociation, the 2'-methoxy spectra show no evidence of nucleobase loss. The w-ions were detected with low abundance. It is known from previous research that RNA has slightly more stability than DNA based on the electronegative 2'-hydroxy group. The reason for the hindered nucleobase loss could be the presence of the electron withdrawing 2'-substituent, which stabilizes the adjacent N-glycosidic bond. Even though the stabilizing effect is not very strong, it is sufficient to reduce the abundance of base loss induced DNA typical [a-B]- and w-ions.

2'-fluoro

Experiments with 2'-fluoro modified oligonucleotides were performed to analyze the stabilizing effect of an electronegative 2'-substituent on the N-glycosidic bond. Results demonstrate that the 2'-fluoro modification significantly reduces the nucleobase loss. Therefore, [a-B]- and w-ions are detected with very low intensity. All other fragments due to backbone cleavage were also observed, but, as in the case of 2'-methoxy oligonucleotides, none is preferred. The experiments suggest that an electronegative 2'-substituent is the reason for the decreased nucleobase loss and consequently also the cause of the changed fragmentation pattern.

Relevance for the dissociation mechanism

RNA, 2'-methoxy and 2'-fluoro oligonucleotides all bear an N-glycosidic bond stabilizing 2'-substituent. This results in low abundances of ions induced by loss of the nucleobase. However, the spectra of RNA compared with those of other 2'-modified oligonucleotides look different. This means that the electronegative substituent only explains the reduced intensity of [a-B]- and w-ions, but not the RNA typical cleavage of the 5'-P-O bond! Thus, there must be another reason for the preferred generation of c- and y-ions in RNA. Logically the hydrogen of the 2'-hydroxyl group must be also involved in the gas-phase dissociation mechanism of RNA. In contrast to 2'-methoxy and 2'-fluoro, the 2'-hydroxyl group of natural RNA provides a weakly bound hydrogen atom, which can build hydrogen bonds and possibly be transferred to a proton acceptor. Subsequently the deprotonated 2'-oxygen could act as a nucleophile, generating a cyclic transition-state.

With the knowledge that the hydrogen of the 2'-hydroxyl group plays a key role in the backbone dissociation of RNA several possible mechanisms are considered. Finally, two mechanisms (JASMS, Scheme 2., p. 35) which both involve the mobile proton were chosen for detailed analyses.

Base modifications

The two proposed dissociation mechanisms of RNA differ in the participation of the nucleobase. Therefore, the role of the nucleobase in the dissociation mechanism was investigated in detail.

dSpacer

The simplest way to test whether the nucleobase is involved in the dissociation mechanism of RNA is a specific modification which should influence the mechanism. In the proposed mechanism, similar to the one known for backbone cleavage in solution, the nucleobase acts as a proton acceptor for the nearby sterical mobile proton of the 5'-situated ribose. (JASMS, Scheme 2b., p. 35). To stop this mechanism, a spacer, a building block lacking the nucleobase, was introduced in the RNA sequence. In the tetramer selected, the third nucleobase was substituted to analyze the backbone cleavage of the adjacent second phosphodiester group. As it was not possible to buy or synthesize a RNA spacer, a DNA spacer was chosen. As explained above the phosphodiester cleavage is only locally influenced. So the deoxyribose leads to the DNA typical fragmentation behavior of the third phosphodiester group. But it has no influence on the observed second phosphodiester group. The lack of the nucleobase should result in a significantly changed fragmentation pattern if this mechanism takes place. Experimental data demonstrate that there is no alternation of the RNA typical fragmentation pattern. Consequently, this means that the nucleobase is not involved in the gas-phase dissociation mechanism of RNA. The confutation of the mechanism involving the nucleobase supports the other proposed mechanism in which the nucleobase does not participate in the RNA dissociation.

Biphenyl oligonucleotides

Experiments with biphenyl-modified DNA and RNA confirm several different aspects of gas-phase dissociation of oligonucleotides. The alcylic biphenyl modifications are C-glycosidic bound. This prevents the base loss.

On the one hand, the spectra of the biphenyl-modified oligoribonucleotides demonstrate unambiguously the lack of dependence of the RNA dissociation mechanism on the nucleobase as a protonation site. The typical c- and y-ions were observed with high abundance independent of an adjacent natural nucleobase or an alcylic biphenyl building block. In contrast to natural nucleobases, the biphenyl can not act as a proton acceptor. Because of that, these results confirm the conclusions reached from dSpacer experiments, which suggest that there is no influence by the nucleobase on the dissociation mechanism of RNA.

On the other hand, experiments with biphenyl-modified DNA show impressively the dependence of DNA dissociation on nucleobase loss as an initiation step. The C-glycosidic bond hinders the naturally occurring cleavage of the N-glycosidic bond in unmodified DNA. Therefore, a loss of the nucleobase is impossible. As expected this leads to the absence of w-ions.

In addition, the lack of backbone cleavage adjacent to the biphenyl (Figure 9.) is an example of the localization of a nucleobase modification within the DNA sequence. However, RNA containing the same modification shows no altered fragmentation pattern. Considering the known DNA and RNA dissociation mechanisms these results are coherent. But they demonstrate that a fundamental understanding of the gas-phase dissociation mechanisms is needed to analyze efficiently modified oligonucleotide spectra.

Conclusions

Experiments with C-glycosidic bound biphenyl-modified oligonucleotides elucidate the influence of the nucleobase on the gas-phase dissociation mechanism. The preferred CID products of natural DNA fragmentation are [a-Base]- and w-ions, which result from the cleavage of the 3'-C-O bond. However, adjacent to the biphenyl modification no cleavage of the phosphodiester backbone was observed, demonstrating clearly the dependence of DNA backbone dissociation on nucleobase loss. RNA fragmentation is characterized by the cleavage of the 5'-P-O bond generating c- and y-ions. No difference in the fragmentation pattern was found between natural RNA and biphenyl-modified building blocks. The results show unambiguously that the gas-phase dissociation of RNA is independent of nucleobase loss.

Experiments with RNA and 2'-modified oligoribonucleotides point out the importance of the 2'-hydroxyl group as the structural key element in the gas-phase dissociation mechanism of RNA. An electronegative 2'-substituent has a stabilizing effect on the N-glycosidic. This is reflected in reduced nucleobase loss. In addition, the availability of the 2'-hydroxyl proton in the steric proximity of the phosphate group is essential for RNA dissociation. Investigations of oligonucleotides containing a spacer demonstrate that the nucleobase is not involved in the RNA dissociation mechanism. These results support the proposed mechanism for RNA gas-phase dissociation.

Fundamental understanding of the dissociation mechanisms of oligonucleotides is needed for the development of tandem mass spectrometric based sequencing methods to be applied to modified oligonucleotides. New methods will facilitate sequence verification in combination with the characterization and localization of modified oligonucleotides used in therapeutic applications.

Outlook

Mechanistic investigations of oligonucleotide dissociation will give way to the characterization of modified oligonucleotides. The focus will be on modifications used in medicine with the aim to develop new methods for fast and reliable sequence verification and detailed structural analyses. However, there are also several other interesting aspects of oligonucleotide dissociation, which deserve attention. For example, there is the elucidation of the positional selectivity of nucleobase loss in DNA or the investigation of the effects of a located charge on backbone dissociation. In addition, the dissociation mechanism of terminal DNA fragmentation is not fully understood and also an alternative fragmentation mechanism responsible for the enhanced formation of w-ions, which is coupled with nucleobase loss, is possible.

In the last few years interesting results demonstrating the localization of metal ions in metal-oligodeoxyribonucleotide complexes were published. Based on the new fundamental information about RNA dissociation, studies of metal-RNA complexes could provide promising information about the characterization of naturally occurring complexes or about the investigation of metal ion induced conformational changes.

IV. References

References

1. Felsenfeld,G. & Rich,A. Studies on the Formation of 2-Stranded and 3-Stranded Polyribonucleotides. *Biochimica et Biophysica Acta* **26**, 457-468 (1957).
2. Felsenfeld,G., Davies,D.R. & Rich,A. Formation of A 3-Stranded Polynucleotide Molecule. *Journal of the American Chemical Society* **79**, 2023-2024 (1957).
3. Ledoan,T., Perrouault,L., Chassignol,M., Thuong,N.T. & Helene,C. Sequence-Targeted Chemical Modifications of Nucleic-Acids by Complementary Oligonucleotides Covalently Linked to Porphyrins. *Nucleic Acids Research* **15**, 8643-8659 (1987).
4. Moser,H.E. & Dervan,P.B. Sequence-Specific Cleavage of Double Helical DNA by Triple Helix Formation. *Science* **238**, 645-650 (1987).
5. Micklefield,J. Backbone modification of nucleic acids: Synthesis, structure and therapeutic applications. *Current Medicinal Chemistry* **8**, 1157-1179 (2001).
6. Buchini,S. & Leumann,C.J. Recent improvements in antigene technology. *Current Opinion in Chemical Biology* **7**, 717-726 (2003).
7. Thuong,N.T. & Helene,C. Sequence-Specific Recognition and Modification of Double-Helical DNA by Oligonucleotides. *Angewandte Chemie-International Edition in English* **32**, 666-690 (1993).
8. Giovannangeli,C. et al. Accessibility of nuclear DNA to triplex-forming oligonucleotides: The integrated HIV-1 provirus as a target. *Proceedings of the National Academy of Sciences of the United States of America* **94**, 79-84 (1997).
9. Wengel,J. Synthesis of 3'-C- and 4'-C-branched oligodeoxynucleotides and the development of locked nucleic acid (LNA). *Accounts of Chemical Research* **32**, 301-310 (1999).
10. Goodchild,J. Oligonucleotide therapeutics: 25 years agrowing. *Current Opinion in Molecular Therapeutics* **6**, 120-128 (2004).
11. Myers,K.J. & Dean,N.M. Sensible use of antisense: how to use oligonucleotides as research tools. *Trends in Pharmacological Sciences* **21**, 19-23 (2000).
12. Baker,B.F. & Monia,B.P. Novel mechanisms for antisense-mediated regulation of gene expression. *Biochim. Biophys. Acta* **1489**, 3-18 (1999).
13. Zamecnik,P.C. & Stephenson,M.L. Inhibition of Rous-Sarcoma Virus-Replication and Cell Transformation by A Specific Oligodeoxynucleotide. *Proceedings of*

the National Academy of Sciences of the United States of America **75**, 280-284 (1978).
14. Stephenson,M.L. & Zamecnik,P.C. Inhibition of Rous-Sarcoma Viral-RNA Translation by A Specific Oligodeoxyribonucleotide. *Proceedings of the National Academy of Sciences of the United States of America* **75**, 285-288 (1978).
15. Torrence,P.F. et al. Recruiting the 2-5A system for antisense therapeutics. *Antisense & Nucleic Acid Drug Development* **7**, 203-206 (1997).
16. Werner,M., Rosa,E., Nordstrom,J.L., Goldberg,A.R. & George,S.T. Short oligonucleotides as external guide sequences for site-specific cleavage of RNA molecules with human RNase P. *RNA-A Publication of the RNA Society* **4**, 847-855 (1998).
17. Kraus,G. et al. Cross-clade inhibition of HIV-1 replication and cytopathology by using RNase P-associated external guide sequences. *Proceedings of the National Academy of Sciences of the United States of America* **99**, 3406-3411 (2002).
18. Dudley,N.R. & Goldstein,B. RNA interference: Silencing in the cytoplasm and nucleus. *Current Opinion in Molecular Therapeutics* **5**, 113-117 (2003).
19. Goodchild,J. Hammerhead ribozymes: Biochemical and chemical considerations. *Current Opinion in Molecular Therapeutics* **2**, 272-281 (2000).
20. Taylor,M.F., Paulauskis,J.D., Weller,D.D. & Kobzik,L. In vitro efficacy of morpholino-modified antisense oligomers directed against tumor necrosis factor-alpha mRNA. *Journal of Biological Chemistry* **271**, 17445-17452 (1996).
21. Summerton,J. & Weller,D. Morpholino antisense oligomers: Design, preparation, and properties. *Antisense & Nucleic Acid Drug Development* **7**, 187-195 (1997).
22. Suwanmanee,T. et al. Restoration of human beta-globin gene expression in murine and human IVS2-654 thalassemic erythroid cells by free uptake of antisense oligonucleotides. *Molecular Pharmacology* **62**, 545-553 (2002).
23. Iversen,P.L. Phosphorodiamidate morpholino oligomers: Favorable properties for sequence-specific gene inactivation. *Current Opinion in Molecular Therapeutics* **3**, 235-238 (2001).
24. Renneberg,D., Bouliong,E., Reber,U., Schumperli,D. & Leumann,C.J. Antisense properties of tricyclo-DNA. *Nucleic Acids Research* **30**, 2751-2757 (2002).

25. Johansson,H.E., Belsham,G.J., Sproat,B.S. & Hentze,M.W. Target-Specific Arrest of Messenger-RNA Translation by Antisense 2'-O-Alkyloligoribonucleotides. *Nucleic Acids Research* **22**, 4591-4598 (1994).
26. Nielsen,P.E. Antisense peptide nucleic acids. *Current Opinion in Molecular Therapeutics* **2**, 282-287 (2000).
27. Knudsen,H. & Nielsen,P.E. Antisense properties of duplex- and triplex-forming PNAs. *Nucleic Acids Research* **24**, 494-500 (1996).
28. Dominski,Z. & Kole,R. Restoration of Correct Splicing in Thalassemic Premessenger RNA by Antisense Oligonucleotides. *Proceedings of the National Academy of Sciences of the United States of America* **90**, 8673-8677 (1993).
29. Sierakowska,H., Sambade,M.J., Agrawal,S. & Kole,R. Repair of thalassemic human beta-globin mRNA in mammalian cells by antisense oligonucleotides. *Proceedings of the National Academy of Sciences of the United States of America* **93**, 12840-12844 (1996).
30. Ruckman,J. et al. 2'-fluoropyrimidine RNA-based aptamers to the 165-amino acid form of vascular endothelial growth factor (VEGF(165)) - Inhibition of receptor binding and VEGF-induced vascular permeability through interactions requiring the exon 7-encoded domain. *Journal of Biological Chemistry* **273**, 20556-20567 (1998).
31. Gold,L., Polisky,B., Uhlenbeck,O. & Yarus,M. Diversity of Oligonucleotide Functions. *Annual Review of Biochemistry* **64**, 763-797 (1995).
32. Osborne,S.E. & Ellington,A.D. Nucleic Acid Selection and the Challenge of Combinatorial Chemistry. *Chem. Rev.* **97**, 349-370 (1997).
33. Lee,M. & Walt,D.R. A fiber-optic microarray biosensor using aptamers as receptors. *Analytical Biochemistry* **282**, 142-146 (2000).
34. Drolet,D.W., MoonMcDermott,L. & Romig,T.S. An enzyme-linked oligonucleotide assay. *Nature Biotechnology* **14**, 1021-1025 (1996).
35. Tuerk,C. & Gold,L. Systematic Evolution of Ligands by Exponential Enrichment - RNA Ligands to Bacteriophage-T4 DNA-Polymerase. *Science* **249**, 505-510 (1990).
36. Bunka,D.H.J. & Stockley,P.G. Aptamers come of age - at last. *Nature Reviews Microbiology* **4**, 588-596 (2006).
37. Varma,R.S. Synthesis of Oligonucleotide Analogs with Modified Backbones. *Synlett* 621-637 (1993).

38. Demesmaeker,A., Altmann,K.H., Waldner,A. & Wendeborn,S. Backbone Modifications in Oligonucleotides and Peptide Nucleic-Acid Systems. *Current Opinion in Structural Biology* **5**, 343-355 (1995).
39. Altmann,K.H. *et al.* Second generation of antisense oligonucleotides: From nuclease resistance to biological efficacy in animals. *Chimia* **50**, 168-176 (1996).
40. Lesnik,E.A. & Freier,S.M. Relative Thermodynamic Stability of DNA, RNA, and DNA-RNA Hybrid Duplexes - Relationship with Base Composition and Structure. *Biochemistry* **34**, 10807-10815 (1995).
41. Verma,S. & Eckstein,F. Modified oligonucleotides: Synthesis and strategy for users. *Annual Review of Biochemistry* **67**, 99-134 (1998).
42. Luyten,I. & Herdewijn,P. Hybridization properties of base-modified oligonucleotides within the double and triple helix motif. *European Journal of Medicinal Chemistry* **33**, 515-576 (1998).
43. Sanghvi,Y.S., Ross,B., Bharadwaj,R. & Vasseur,J.J. An Easy Access of 2',3'-Dideoxy-3'-Alpha-C-Formyl-Adenosine and 2',3'-Dideoxy-3'-Alpha-C-Formyl-Guanosine Analogs Via Stereoselective C-C Bond-Forming Radical Reaction. *Tetrahedron Letters* **35**, 4697-4700 (1994).
44. Mcluckey,S.A., Vanberkel,G.J. & Glish,G.L. Tandem Mass-Spectrometry of Small, Multiply Charged Oligonucleotides. *Journal of the American Society for Mass Spectrometry* **3**, 60-70 (1992).
45. Mcluckey,S.A. & Habibigoudarzi,S. Decompositions of Multiply-Charged Oligonucleotide Anions. *Journal of the American Chemical Society* **115**, 12085-12095 (1993).
46. Rodgers,M.T., Campbell,S., Marzluff,E.M. & Beauchamp,J.L. Low-Energy Collision-Induced Dissociation of Deprotonated Dinucleotides - Determination of the Energetically Favored Dissociation Pathways and the Relative Acidities of the Nucleic-Acid Bases. *International Journal of Mass Spectrometry and Ion Processes* **137**, 121-149 (1994).
47. Barry,J.P., Vouros,P., Vanschepdael,A. & Law,S.J. Mass and Sequence Verification of Modified Oligonucleotides Using Electrospray Tandem Mass-Spectrometry. *Journal of Mass Spectrometry* **30**, 993-1006 (1995).
48. Wu,J. & Mcluckey,S.A. Gas-phase fragmentation of oligonucleotide ions. *International Journal of Mass Spectrometry* **237**, 197-241 (2004).

49. Wan,K.X., Gross,J., Hillenkamp,F. & Gross,M.L. Fragmentation mechanisms of oligodeoxynucleotides studied by H/D exchange and electrospray ionization tandem mass spectrometry. *Journal of the American Society for Mass Spectrometry* **12**, 193-205 (2001).

50. Wang,Z., Wan,K.X., Ramanathan,R., Taylor,J.S. & Gross,M.L. Structure and fragmentation mechanisms of isomeric T-rich oligodeoxynucleotides: A comparison of four tandem mass spectrometric methods. *Journal of the American Society for Mass Spectrometry* **9**, 683-691 (1998).

51. Pan,S., Verhoeven,K. & Lee,J.K. Investigation of the initial fragmentation of oligodeoxynucleotides in a quadrupole ion trap: Charge level-related base loss. *Journal of the American Society for Mass Spectrometry* **16**, 1853-1865 (2005).

52. Mcluckey,S.A., Vaidyanathan,G. & Habibigoudarzi,S. Charged Vs Neutral Nucleobase Loss from Multiply-Charged Oligonucleotide Anions. *Journal of Mass Spectrometry* **30**, 1222-1229 (1995).

53. Mcluckey,S.A. & Vaidyanathan,G. Charge state effects in the decompositions of single-nucleobase oligonucleotide polyanions. *International Journal of Mass Spectrometry and Ion Processes* **162**, 1-16 (1997).

54. Daneshfar,R. & Klassen,J.S. Arrhenius activation parameters for the loss of neutral nucleobases from deprotonated oligonucleotide anions in the gas phase. *Journal of the American Society for Mass Spectrometry* **15**, 55-64 (2004).

55. Wan,K.X. & Gross,M.L. Fragmentation mechanisms of oligodeoxynucleotides: Effects of replacing phosphates with methylphosphonates and thymines with other bases in T-rich sequences. *Journal of the American Society for Mass Spectrometry* **12**, 580-589 (2001).

56. Little,D.P., Aaserud,D.J., Valaskovic,G.A. & McLafferty,F.W. Sequence information from 42-108-mer DNAs (complete for a 50-mer) by tandem mass spectrometry. *Journal of the American Chemical Society* **118**, 9352-9359 (1996).

57. Luo,H., Lipton,M.S. & Smith,R.D. Charge effects for differentiation of oligodeoxynucleotide isomers containing 8-oxo-dG residues. *Journal of the American Society for Mass Spectrometry* **13**, 195-199 (2002).

58. SannesLowery,K.A., Mack,D.P., Hu,P.F., Mei,H.Y. & Loo,J.A. Positive ion electrospray ionization mass spectrometry of oligonucleotides. *Journal of the American Society for Mass Spectrometry* **8**, 90-95 (1997).

59. Gross,J. et al. Investigations of the metastable decay of DNA under ultraviolet matrix-assisted laser desorption/ionization conditions with post-source-decay analysis and hydrogen/deuterium exchange. *Journal of the American Society for Mass Spectrometry* **9**, 866-878 (1998).
60. Liu,D.F., Wyttenbach,T. & Bowers,M.T. Hydration of mononucleotides. *Journal of the American Chemical Society* **128**, 15155-15163 (2006).
61. Tang,W., Zhu,L. & Smith,L.M. Controlling DNA fragmentation in MALDI-MS by chemical modification. *Analytical Chemistry* **69**, 302-312 (1997).
62. Nordhoff,E. et al. Ion Stability of Nucleic-Acids in Infrared Matrix-Assisted Laser-Desorption Ionization Mass-Spectrometry. *Nucleic Acids Research* **21**, 3347-3357 (1993).
63. Cerny,R.L., Tomer,K.B., Gross,M.L. & Grotjahn,L. Fast-Atom-Bombardment Combined with Tandem Mass-Spectrometry for Determining Structures of Small Oligonucleotides. *Analytical Biochemistry* **165**, 175-182 (1987).
64. Kirpekar,F. & Krogh,T.N. RNA fragmentation studied in a matrix-assisted laser desorption/ionisation tandem quadrupole/orthogonal time-of-flight mass spectrometer. *Rapid Communications in Mass Spectrometry* **15**, 8-14 (2001).
65. Schurch,S., Bernal-Mendez,E. & Leumann,C.J. Electrospray tandem mass spectrometry of mixed-sequence RNA/DNA oligonucleotides. *Journal of the American Society for Mass Spectrometry* **13**, 936-945 (2002).
66. Andersen,T.E., Kirpekar,F. & Haselmann,K.F. RNA fragmentation in MALDI mass spectrometry studied by H/D-exchange: Mechanisms of general applicability to nucleic acids. *Journal of the American Society for Mass Spectrometry* **17**, 1353-1368 (2006).
67. Maxam,A.M. & Gilbert,W. New Method for Sequencing DNA. *Proceedings of the National Academy of Sciences of the United States of America* **74**, 560-564 (1977).
68. Sanger,F., Nicklen,S. & Coulson,A.R. DNA Sequencing with Chain-Terminating Inhibitors. *Proceedings of the National Academy of Sciences of the United States of America* **74**, 5463-5467 (1977).
69. Cerny,R.L., Gross,M.L. & Grotjahn,L. Fast-Atom-Bombardment Combined with Tandem Mass-Spectrometry for the Study of Dinucleotides. *Analytical Biochemistry* **156**, 424-435 (1986).

70. Little,D.P. et al. Rapid Sequencing of Oligonucleotides by High-Resolution Mass-Spectrometry. *Journal of the American Chemical Society* **116**, 4893-4897 (1994).
71. Oberacher,H., Parson,W., Oefner,P.J., Mayr,B.M. & Huber,C.G. Applicability of tandem mass spectrometry to the automated comparative sequencing of long-chain oligonucleotides. *Journal of the American Society for Mass Spectrometry* **15**, 510-522 (2004).
72. Oberacher,H., Mayr,B.M. & Huber,C.G. Automated de novo sequencing of nucleic acids by liquid chromatography-tandem mass Spectrometry. *Journal of the American Society for Mass Spectrometry* **15**, 32-42 (2004).
73. Oberacher,H., Wellenzohn,B. & Huber,C.G. Comparative sequencing of nucleic acids by liquid chromatography-tandem mass spectrometry. *Analytical Chemistry* **74**, 211-218 (2002).
74. Hahn,C.S., Strauss,E.G. & Strauss,J.H. Dideoxy sequencing of RNA using reverse transcriptase. *Methods Enzymol.* **180**, 121-130 (1989).
75. Thomas,B. & Akoulitchev,A.V. Mass spectrometry of RNA. *Trends in Biochemical Sciences* **31**, 173-181 (2006).
76. Wang,J. et al. Cyclohexene nucleic acids (CeNA): Serum stable oligonucleotides that activate RNase H and increase duplex stability with complementary RNA. *Journal of the American Chemical Society* **122**, 8595-8602 (2000).
77. Qiu,F. & McCloskey,J.A. Selective detection of ribose-methylated nucleotides in RNA by a mass spectrometry-based method. *Nucleic Acids Res.* **27**, e20 (1999).
78. Pomerantz,S.C. & McCloskey,J.A. Detection of the common RNA nucleoside pseudouridine in mixtures of oligonucleotides by mass spectrometry. *Analytical Chemistry* **77**, 4687-4697 (2005).
79. Bartlett,M.G., McCloskey,J.A., Manalili,S. & Griffey,R.H. The effect of backbone charge on the collision-induced dissociation of oligonucleotides. *Journal of Mass Spectrometry* **31**, 1277-1283 (1996).
80. Sannes-Lowery,K.A. & Hofstadler,S.A. Sequence confirmation of modified oligonucleotides using IRMPD in the external ion reservoir of an electrospray ionization Fourier transform ion cyclotron mass spectrometer. *Journal of the American Society for Mass Spectrometry* **14**, 825-833 (2003).
81. Kawasaki,A.M. et al. Uniformly Modified 2'-Deoxy-2'-Fluoro Phosphorothioate Oligonucleotides As Nuclease-Resistant Antisense Compounds with High-Affinity

and Specificity for RNA Targets. *Journal of Medicinal Chemistry* **36**, 831-841 (1993).

82. Polo,L.M., McCarley,T.D. & Limbach,P.A. Chemical sequencing of phosphorothioate oligonucleotides using matrix-assisted laser desorption ionization time-of-flight mass spectrometry. *Analytical Chemistry* **69**, 1107-1112 (1997).

I want morebooks!

Buy your books fast and straightforward online - at one of the world's fastest growing online book stores! Environmentally sound due to Print-on-Demand technologies.

Buy your books online at

www.get-morebooks.com

Kaufen Sie Ihre Bücher schnell und unkompliziert online – auf einer der am schnellsten wachsenden Buchhandelsplattformen weltweit!
Dank Print-On-Demand umwelt- und ressourcenschonend produziert.

Bücher schneller online kaufen
www.morebooks.de

OmniScriptum Marketing DEU GmbH
Heinrich-Böcking-Str. 6-8
D - 66121 Saarbrücken
Telefax: +49 681 93 81 567-9

info@omniscriptum.com
www.omniscriptum.com

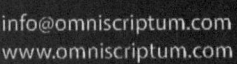

Printed by Books on Demand GmbH, Norderstedt / Germany